T0256063

**Computational Modelling and
Simulation of Aircraft
and the Environment**

Aerospace Series

Visit www.wiley.com to view more titles in the Aerospace Series.

Computational Modelling and Simulation of Aircraft and the Environment

Volume 2: Aircraft Dynamics

Dominic J. Diston
University of Liverpool
UK

Registered Offices
John Wiley & Sons, Inc., 111 River Street, Hoboken, NJ 07030, USA
John Wiley & Sons Ltd, The Atrium, Southern Gate, Chichester, West Sussex, PO19 8SQ, UK

For details of our global editorial offices, customer services, and more information about Wiley products visit us at www.wiley.com.

Wiley also publishes its books in a variety of electronic formats and by print-on-demand. Some content that appears in standard print versions of this book may not be available in other formats.

Library of Congress Cataloging-in-Publication Data Applied for:

Hardback ISBN: 9780470687116

Cover Design: Wiley
Cover Image: © Andrei Armiagov/Shutterstock

Set in 9.5/12.5pt STIXTwoText by Straive, Pondicherry, India
Printed and bound by CPI Group (UK) Ltd, Croydon, CR0 4YY

C9780470687116_010224

Contents

Preface

Many years have passed since the publication of Volume 1, far more than ever intended or expected. This has been an inescapable consequence of a high workload elsewhere and, until recently, a general shortage of time caused by work/life imbalance. However, here it is ... a range of material that covers some of the stuff that I have dealt with in a career that spanned approximately 43 years. In the 24 years spent at BAE SYSTEMS (and previously British Aerospace), I had the privilege of working on some of the most interesting and inventive areas of engineering that I could have ever wished for, as well as the privilege of working with some extraordinarily talented people. In the 19 years spent in higher education, I taught a wide range of aerospace subjects at the Universities of Manchester, Liverpool and Nottingham, as well as a short period at the Empire Test Pilots' School. So, this book is a digest of subject matter in the area of vehicle modelling and simulation. It focuses on the flight physics of fixed-wing aircraft, with a short introduction towards the end of the book on systems modelling. This is all based on project work and taught modules that I have produced during my career. I have attempted to be thorough and interesting, although readers might not necessarily agree. However, it should provide more than enough information and interpretation to help develop specialised knowledge. I do not guarantee perfection, but I did employ my best endeavours to make this a worthwhile publication, albeit later than planned.

Aerospace Series Preface

The field of aerospace is multi-disciplinary and wide ranging, covering a large variety of products, disciplines and domains, not merely in engineering but in many related supporting activities. The combination of these elements enables the aerospace industry to produce innovative and technologically advanced vehicles and systems. The wealth of knowledge and experience that has been gained by expert practitioners in the various aerospace fields needs to be passed onto others working in the industry as well as to researchers, teachers and the student body in universities.

The *Aerospace Series* aims to be a practical, topical and relevant series of books aimed at people working in the aerospace industry, including engineering professionals and operators, engineering educators in academia, and allied professions such as commercial and legal executives. The range of topics is intended to be wide, covering design and development, manufacture, operation and support of aircraft, as well as infrastructure operations and current advances in research and technology.

In the field of aircraft systems design, the application of computational models using commercially available modelling and simulation tools is widely practiced and taught in most university engineering degree courses. The use of such models provides the system designer with the ability to rapidly model their preliminary design concept and to exercise the model under many operational and failure scenarios. This approach allows the performance of the system to be assessed and enables the introduction of changes as required to refine the design. In this way, it is possible to validate the behaviour of the system before committing to the final detailed design – a valuable contribution to 'right first time' design.

Computational Modelling of Aircraft and the Environment, Volume 2 : Aircraft Dynamics is the sequel to Volume 1 which focused upon *Platform Kinetics and Synthetic Environments* and introduced a means of modelling the real-world environment in which aircraft operate. This book continues the journey of providing a broad-based text covering applicable mathematics and science in key domains required to create models to assist with the design of systems by summarising the essential elements of air vehicle modelling and simulation with a focus on the flight physics of fixed-wing aircraft. It also introduces the development of representative flight models, deriving the equations of motion, fixed-wing aerodynamics, longitudinal flight mechanics and the fundamentals of gas turbine dynamics. The final section builds upon the previous material through consideration of structural models, mass properties and physical system modelling, including the use of bond graphs. This book is a welcome addition to the Wiley Aerospace Series.

Peter Belobaba, Jonathan Cooper and Allan Seabridge
December 2023

1

A Simple Flight Model

1.1 Introduction

1.1.1 General Introduction to Volume 2

Welcome to Volume 2 of *Computational Modelling and Simulation of Aircraft and the Environment*. This volume will present and explain the main theories that enable the dynamics of fixed-wing aircraft to be modelled using mathematical and computational methods. The aim is to establish the heuristic basis for education in aeronautical engineering that provides a 'handbook' of concepts and interpretations, together with a formulary to support practical application. It is appropriate and convenient to commence with a simple flight model that brings together all the essential components without too much detail. This covers aircraft motion, atmosphere, aerodynamics, and propulsion. More detailed expositions are given in Chapters 2–5. These focus on Equations of Motion, Wing Aerodynamics, Longitudinal Flight and Gas Turbines.

The significant omission is lateral-directional aerodynamics, apart from rolling a wing in flight (later in Chapter 1). This is because the formulary tends to be complicated and abstract, with no easily recognisable link to the underlying physics. Also, there is no inherent value in just repeating what other books [e.g. Pamadi] already provide. Also, supersonic flight is not discussed because it is a specialised area of aircraft design. The vast majority of aircraft are not supersonic.

The final chapter offers a brief introduction to several topics that are important in whole-aircraft modelling but that sit outside the usual scope of flight physics. The discussion is brief because these subjects have substantial content and could easily expand to fill another two or three textbooks.

1.1.2 What Chapter 1 Includes

This chapter includes:

- Equations of motions expressed with respect to flight path parameters.
- Summary of the International Standard Atmosphere up to 20 km (roughly 50 000 ft).
- Simple propulsion model that enables thrust calculations at given altitude and Mach number.

Computational Modelling and Simulation of Aircraft and the Environment: Aircraft Dynamics, First Edition, Volume II. Dominic J. Diston.
© 2024 John Wiley & Sons Ltd. Published 2024 by John Wiley & Sons Ltd.

- Simple aerodynamic model that is applicable to idealised wing geometry plus trailing-edge flaps.
- A short introduction to spanwise lift distribution for an idealised wing.
- Aerodynamic model for wing/tail combinations (as an approximation to a complete aircraft).
- A set of airspeed definitions.
- One of many possible architectures for a flight model (i.e. a whole-aircraft model).

1.1.3 What Chapter 1 Excludes

This chapter excludes:

- Six degree-of-freedom (6-DOF) equations of motion [go to Chapter 2].
- Generalised wing configurations (e.g. taper, twist) [go to Chapter 3].
- Flight mechanics of wing/tail combinations [go to Chapter 4].
- Fuselage aerodynamic effects [go to Chapter 4].
- Physics-based models of gas turbines [go to Chapter 5].
- Lateral-directional aerodynamics [not covered by this book].
- Supersonic flight [not covered by this book].

1.1.4 Overall Aim

Chapter 1 should provide 'enough of everything' that is needed to create a complete representation of aircraft flight behaviour, from ground up to 20 km and from low-speed up to about 0.85 Mach number. This includes the essential flight physics without too much detail, such that computations can be verified by manual calculation and that parametric trend should be readily discernible. In short, this should provide a compact aircraft model for the purpose of preliminary concept evaluation and simulation.

1.2 Flight Path

The simplest possible flight path model is shown in Figure 1.1. This represents symmetric flight (with wings level) in a vertical plane. Motion parameters are defined at the centre of mass for an instantaneous pull-up (which is turn in the vertical plane). Airspeed V is aligned (or tangential) with the flight path, which is normal to the radius of turn. The tangential acceleration varies the airspeed while the centripetal acceleration varies the flight path angle. The pitch angle θ defines the orientation of the aircraft horizontal datum and the angle of attack (AOA) is defined by:

$$\alpha = \theta - \gamma \qquad (1.1)$$

where γ is the climb/dive angle. The rate of change of pitch angle is the pitch rate $\dot{\theta} = q$, such that

$$\dot{\alpha} = q - \dot{\gamma} \qquad (1.2)$$

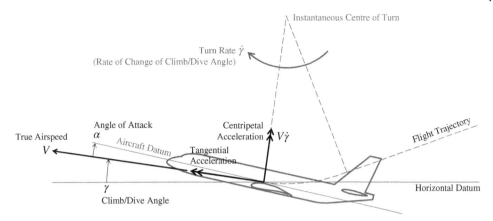

Figure 1.1 Symmetric Flight Trajectory.

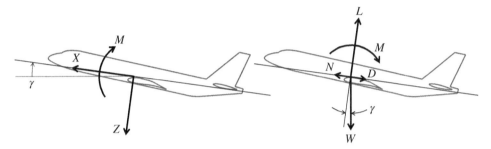

Figure 1.2 Symmetric Force/Moment System.

The force/moment system is shown in Figure 1.2 (referred to the centre of gravity, CG). Thus, aircraft motion is governed by the following equations when aircraft mass is constant:

$$m\dot{V} = X \qquad mV\dot{\gamma} = -Z \qquad J\dot{q} = M \qquad (1.3)$$

where m is aircraft mass, J is moment of inertia, X is tangential force, Z is normal force and M is pitching moment. Altenatively, these equations can be written as:

$$\dot{V} = \frac{X}{m} \qquad \dot{\gamma} = -\frac{Z}{mV} \qquad \dot{\alpha} = \frac{M}{J} - \dot{\gamma} \qquad (1.4)$$

The forces X and Z are composed as:

$$X = T - D - W\sin\gamma \qquad Z = W\cos\gamma - L \qquad (1.5)$$

where L is total lift, D is total drag, W is aircraft weight and T is the total nett thrust from all engines. For convenience, the thrust line is drawn through the centre of mass. Also, for convenience, the thrust is aligned with the velocity vector and not the aircraft datum. This is true if AOA is zero (which it rarely is) and almost true if AOA is small (which it usually is).

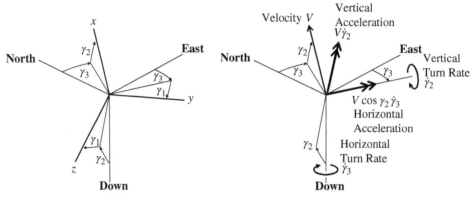

Figure 1.3 Generalised Flight Path Parameters.

Currently, the flight path is constrained to lie within a single vertical plane, tracing a straight line course across the surface of the Earth. Horizontal turns would be useful! So, a reference system is defined for the Earth, with its axes aligned with North, East, and Down, shown in Figure 1.3. Flight path angles are defined as γ_3 (setting the course direction), γ_2 (setting the climb/dive angle), and γ_1 (setting a rotation about the velocity vector). The resulting 'flight path axes' are shown as xyz. The vertical turn rate is now written as $\dot{\gamma}_2$ [cf. Figure 1.1] and a horizontal turn rate is introduced as $\dot{\gamma}_3$. In fact, $\dot{\gamma}$ can be redefined as the variation in flight path angle measured in the plane of symmetry (which is inclined at an angle γ_1 with respect to the vertical):

$$\dot{\gamma} = \dot{\gamma}_2 \cos \gamma_1 + \dot{\gamma}_3 \cos \gamma_2 \sin \gamma_1 \tag{1.6}$$

The force/moment system is modified and extended, as shown in Figure 1.4. The lift vector is inclined at an angle γ_1 with respect to the vertical. This generates the horizontal acceleration, thereby providing a bank-to-turn capability. Rotation about the velocity vector is produced by a rolling moment K about the velocity vector, such that the roll rate p is equal to $\dot{\gamma}_1$.

The generalised equations of motion are given by:

$$\dot{V} = \frac{N-D}{m} - g \sin \gamma_2 \qquad \dot{\gamma}_2 = \frac{L}{mV} \cos \gamma_1 - \frac{g}{V} \cos \gamma_2 \qquad \dot{\gamma}_3 = \frac{L}{mV} \frac{\sin \gamma_1}{\cos \gamma_2} \tag{1.7}$$

$$\dot{p} = \frac{K}{J_1} \qquad \dot{\gamma}_1 = p \qquad \dot{q} = \frac{M}{J_2} \qquad \dot{\alpha} = q - \dot{\gamma} \tag{1.8}$$

where m is the aircraft mass, g is the gravitational acceleration, J_1 is the roll moment of inertia, J_2 is the pitch moment of inertia, and the other symbols have their previously defined meanings.

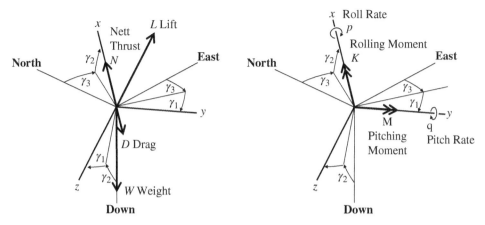

Figure 1.4 Generalised Force/Moment System.

The flight path angles are $(\gamma_1, \gamma_2, \gamma_3)$. In addition, aircraft position (expressed as $e =$ East, $n =$ North, and $h =$ Altitude) is given by simple trigonometry:

$$\dot{e} = V \cos\gamma_2 \sin\gamma_3 \qquad \dot{n} = V \cos\gamma_2 \cos\gamma_3 \qquad \dot{h} = -V \sin\gamma_2 \tag{1.9}$$

1.3 Flight Environment <20 km

Atmosphere models were explained and developed in Volume 1. The International Standard Atmosphere is underpinned by parametric values given in Table 1.1. Temperature T [measured in Kelvin] decreases with altitude up to 11 km (at the top of the troposphere) and remains constant up to 20 km (at the top of the lower stratosphere), such that:

$$T = \max(T_0 + L_{01}H, T_1) \quad \text{for} \quad H \leq 20\,000 \text{ m} \tag{1.10}$$

where H is the geopotential altitude. This altitude scale is used because the variation of pressure P can be expressed with a constant value for gravitational acceleration g_0 (at sea level):

$$dP = -\rho g_0 dH \tag{1.11}$$

where ρ is air density. The relationship with altitude h (from Section 1.2) is as follows:

$$H = \frac{r_0 h}{r_0 + h} \qquad h = \frac{r_0 H}{r_0 - H} \tag{1.12}$$

and where r_0 is the mean radius of the Earth.

The relationship between pressure P and temperature T is given by the Ideal Gas Law:

$$P = \rho RT \tag{1.13}$$

Accordingly, Equation 1.11 gives:

$$\int \frac{dP}{P} = -\frac{g_0}{R} \int \frac{1}{T} dH$$

Table 1.1 Selected Parameters for the International Standard Atmosphere.

Parameter	Symbol	Value	Units
Mean Earth radius	r_0	6356766	m
Gravitational acceleration at sea level	g_0	9.80665	$m\,s^{-2}$
Gas constant (dry air)	R	287.05287	$(J\,kg^{-1})\,K^{-1}$
Adiabatic gas constant (dry air)	γ	1.4	
Pressure at sea level	P_0	101325	Pa
Pressure at Tropopause	P_1	22632	Pa
Temperature at sea level	T_0	288.15	K
Temperature at Tropopause	T_1	216.65	K
Temperature gradient across Troposphere	L_{01}	-0.0065	$K\,m^{-1}$
Density at sea level	$\rho_0 = P_0/RT_0$	1.225	$kg\,m^{-3}$
Speed of sound at sea level	$a_0 = \sqrt{\gamma RT_0}$	340.294	$m\,s^{-1}$

Recalling Equation 1.10, the pressure variation across the troposphere is determined as:

$$\frac{P}{P_0} = \left(\frac{T}{T_0}\right)^{-\frac{g_0}{RL_{01}}} \tag{1.14a}$$

and the pressure variation across the lower stratosphere (at constant temperature) is:

$$\frac{P}{P_1} = \exp\left[-\frac{g_0}{RT_1}(H - 11\,000)\right] \tag{1.14b}$$

Air density ρ is then determined from the Ideal Gas Law.

It is noted that the local speed of sound is defined by $\sqrt{\gamma RT}$ (as discussed in Chapter 5) and, thus, airspeed V can be expressed by its Mach number M_N:

$$M_N = \frac{V}{\sqrt{\gamma RT}} \tag{1.15}$$

1.4 Simple Propulsion Model

1.4.1 Reference Parameters

Aeronautical engineering makes use of equivalent sea-level parameters in order to enable comparisons to be made at different altitudes. It is common to relate pressure, temperature, and density to their respective sea-level values. The resulting quantities are called relative pressure δ, relative temperature θ and relative density σ:

$$\delta = \frac{P}{P_0} \qquad \theta = \frac{T}{T_0} \qquad \sigma = \frac{\rho}{\rho_0} \tag{1.16}$$

where the subscript 0 denotes Sea Level. It is noted that the Ideal Gas Law (cf. Equation 1.13) can now be rewritten as:

$$\delta = \sigma\theta \tag{1.17}$$

Referred Airspeed V_0 gives the same Mach number at sea level as for the airspeed V at a given altitude, such that:

$$\frac{V}{V_0} = \frac{M_N\sqrt{\gamma RT}}{M_N\sqrt{\gamma RT_0}} = \sqrt{\frac{T}{T_0}} = \sqrt{\theta} \tag{1.18}$$

Referred Massflow Q_0 is based on referred airspeed for a given cross-sectional flow area A:

$$\frac{Q}{Q_0} = \frac{\rho AV}{\rho_0 AV_0} = \sigma\sqrt{\theta} = \frac{\delta}{\sqrt{\theta}} \tag{1.19}$$

Referred Thrust X_0 is based on relative pressure applied over a surface area A:

$$\frac{X}{X_0} = \frac{AP}{AP_0} = \delta \tag{1.20}$$

Referred Fuel Flow F_0 (which is energy flow) is based on the momentum force that is calculated as the product of thrust X and airspeed V:

$$\frac{F}{F_0} = \frac{XV}{X_0V_0} = \delta\sqrt{\theta} \tag{1.21}$$

1.4.2 Simple Jet Engine Performance

A simple engine model is presented here that was created so many years ago that the original source document has been lost and the team that produced it has been forgotten. It was used in the 1970s for research into autoland systems and it serves a useful purpose here because of its simplicity.

Engine parameters are Gross Thrust X, Intake Massflow Q, Fuel Massflow F and Engine Speed Ω. These are referred to sea level as follows:

$$X_0 = \frac{X}{\delta} \qquad Q_0 = Q\frac{\sqrt{\theta}}{\delta} \qquad F_0 = \frac{F}{\delta\sqrt{\theta}} \qquad \Omega_0 = \frac{\Omega}{\sqrt{\theta}} \tag{1.22}$$

where δ is relative pressure and θ is relative temperature. Note that engine speed (which is an angular quantity) is treated in the same manner as airspeed (which is a linear quantity). Maximum values of these parameters are X_{max}, Q_{max}, F_{max} and Ω_{max}. The corresponding nondimensional parameters are:

$$x = \frac{X_0}{X_{max}} \qquad q = \frac{Q_0}{Q_{max}} \qquad f = \frac{F_0}{F_{max}} \qquad \omega = \frac{\Omega_0}{\Omega_{max}} \tag{1.23}$$

Engine performance is defined empirically as follows:

$$q = 1.10\omega - 0.10 \tag{1.24}$$

$$x = 1.95\omega^2 - 1.20\omega + 0.25 \tag{1.25}$$

$$f = max((0.60\omega - 0.06), (2.40\omega - 1.40)) \tag{1.26}$$

Equation 1.25 can only be solved for $x > 0.06539$ which corresponds with $\omega > 0.3077$. Nett thrust is the gross thrust minus the intake momentum drag:

$$N = X - QV \tag{1.27}$$

Using referred parameters:

$$\frac{N}{\delta} = X_0 - Q_0 V_0 \tag{1.28}$$

Applying nondimensional parameters:

$$\frac{N}{\delta} = xX_{max} - qQ_{max}a_0 M_N \tag{1.29}$$

where M_N is Mach number and a_0 is the speed of sound at sea level. Maximum nett thrust is given when $x = 1$ and $q = 1$:

$$\frac{N_{max}}{\delta} = X_{max} - Q_{max}a_0 M_N \tag{1.30}$$

Thus, using Equations 1.22 and 1.23, the nett thrust at any flight condition is given by:

$$\frac{N}{\delta} = \left(1.95\omega^2 - 1.20\omega + 0.25\right)X_{max} - \left(1.10\omega - 0.10\right)Q_{max}a_0 M_N \tag{1.31}$$

As an example, assume the following values associated with maximum thrust at sea level:

$$X_{max} = 55\,000 \text{ N} \qquad Q_{max} = 95 \text{ kg} \cdot \text{s}^{-1} \qquad F_{max} = 0.9 \text{ kg} \cdot \text{s}^{-1} \qquad \Omega_{max} = 9\,000 \text{ rpm}$$

This corresponds with a specific fuel consumption of 58.9 $(\text{kg h}^{-1}) \text{ kN}^{-1}$. Nett thrust is given by:

$$\frac{1}{\delta}\left(\frac{N}{55\,000}\right) = 1.95\omega^2 - (1.20 + 0.64658M)\omega + (0.25 + 0.05878M) $$

From this formula, the variation of referred net thrust against referred engine speed and Mach number is calculated and shown in Figure 1.5. The accuracy of this model diminishes significantly as altitude increases but it is adequate for its original purpose, investigating automatic control for low-altitude, low-speed flight.

1.4.3 'Better' Jet Engine Performance

Better thrust estimates are available but, for illustrative purposes, only one will be considered here. The variation with Mach number is given by Bartel and Young (2007), following the method of Torenbeek (1982):

$$\frac{N}{N_{TO}} = 1 - k_1 M + k_2 M^2 \tag{1.32}$$

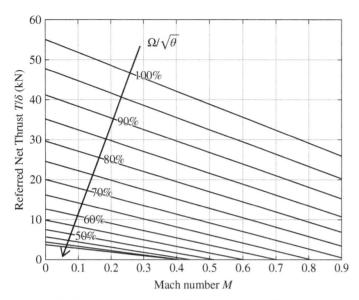

Figure 1.5 Thrust Performance for Simple Jet Engine Model.

where N is nett thrust, N_{TO} is nett thrust at take-off and M is Mach number. The factors k_1 and k_2 were determined from studying a number of two-shaft engines:

$$k_1 = \frac{0.377(1 + \lambda)}{\sqrt{(1 + 0.82\lambda)G}} \qquad k_2 = 0.19\lambda + 0.23 \tag{1.33}$$

where λ is the bypass ratio and G is the so-called gas generator function:

$$G = 0.06\sqrt{\lambda} + 0.64$$

For example, a Rolls-Royce Spey has $\lambda = 0.64$ and so $k_1 = 0.559$ and $k_2 = 0.678$.
Equation 1.32 is extended to give the variation of thrust with Mach number and altitude:

$$\frac{N}{N_{TO}} = A_0 - A_1 k_1 M + A_2 k_2 M^2 \tag{1.34}$$

where the new coefficients are stated as:

$$A_0 = -0.4327\delta^2 + 1.3855\delta + 0.0472$$

$$A_1 = 0.9106\delta^3 - 1.7736\delta^2 + 1.8697\delta$$

$$A_2 = 0.1377\delta^3 - 0.4374\delta^2 + 1.3003\delta$$

in which δ is the relative pressure that was defined in Equation 1.14.

The variation of thrust-specific fuel consumption (TSFC) is given by Howe (2000):

$$c = c_2 \left(1 - 0.15\lambda^{0.65}\right) \left(1 + 0.28\left(1 + 0.063\lambda^2\right)M\right)\sigma^{0.08} \tag{1.35}$$

where c is TSFC, λ is bypass ratio, M is Mach number and σ is relative density (ρ/ρ_0). The factor c_2 is chosen as a constant that matches the performance of a specific engine but, in the absence of real data, it is claimed that preliminary analysis can use $c_2 = 24 \, (\text{mg s}^{-1}) \, \text{N}^{-1}$ for a low-bypass engine and $c_2 = 20$ (mg s^{-1}) N^{-1} for a high-bypass engine. In more practical units, $c_2 = 86.4$ (kg h^{-1}) kN^{-1} for a low-bypass engine and $c_2 = 72$ (kg h^{-1}) kN^{-1} for a high-bypass engine.

1.4.4 Simple Jet Engine Dynamics

The engine speed for a given fuel flow is defined by the inverse of Equation 1.26. The solution can be interpreted as a speed demand:

$$\omega_{demand} = min\left(\frac{f + 0.06}{0.6}, \frac{f + 1.40}{2.40}\right) \tag{1.36}$$

For preliminary calculations, It is appropriate to model engine dynamics as a first-order response to a speed demand, with a time constant (τ) and an acceleration limit (A):

$$\frac{d\omega}{dt} = \frac{1}{\tau}(\omega_{demand} - \omega)\Big]_{-A}^{A} \tag{1.37}$$

As an initial guess, choose a time constant of 0.5 second for small perturbations together with a rate limit that is compatible with an acceleration from 'flight idle' (defined as $\omega = 0.325$) up to maximum speed ($\omega = 1.00$) nominally in 5 seconds.

1.5 Simple Aerodynamic Model

1.5.1 Idealised Aircraft

An idealised aircraft configuration provides the essential constituents for building a model of aircraft flight physics. So, this section will assume a conventional wing/tail combination, such as shown in Figure 1.6. This enables a compact set of aerodynamic calculations to be defined, without becoming buried under an avalanche of empirical detail.

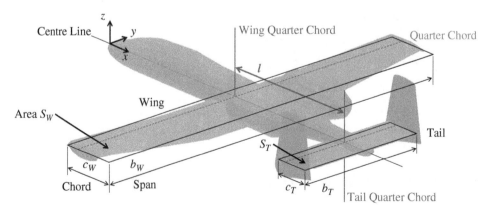

Figure 1.6 Idealised Aircraft.

Geometric axes for airframe definition are defined by the **xyz** triad with its point of origin at the intersection of the aircraft centre line, the horizontal datum and the traverse datum (which is located at the nose in this example). For simplicity, the lifting surfaces have a rectangular planform with span b, chord length c, area $S = bc$ and aspect ratio $A = b/c$. Subscripts W and T denote wing and tail, respectively. Reference points are defined at the quarter-chord positions for the wing and tail, at which aerodynamic forces and moments are calculated. The distance between these points is l. The product of area and distance is the *tail volume* ($V_T = lS_T$), which determines the effectiveness of the tail as a producer of pitching moment. For convenience, the reference points are assumed to lie on the intersection of the aircraft centre line and the horizontal datum, together with the aircraft CG. This simplifies the analysis of flight stability.

1.5.2 Idealised Wing

Figure 1.3 summarises the aerodynamics of a simple wing (generating lift L, drag D, and pitching moment M) for true airspeed V and AOA α. In 'normal' flight, lift variation is linear with respect to AOA and also AOA is reasonably small (probably less than 10° in most conditions). The pitching moment is nominally constant when calculated about the quarter-chord, for reasons that are discussed in Chapter 3. If the wing cross-section is symmetrical (as it is in Figure 1.7), then the pitching moment is zero and, in addition, the wing generates no lift at zero AOA.

Lift and drag are defined as follows:

$$L = QS\,C_L \qquad\qquad D = QS\,C_D \qquad\qquad\qquad (1.38)$$

where the dynamic pressure is $Q = \rho V^2/2$ and S is the wing area. The quantities C_L and C_D are aerodynamic coefficients for lift and drag, respectively. For this simplified (symmetric) wing, the lift coefficient is defined as:

$$C_L = a\alpha \qquad\qquad\qquad\qquad\qquad\qquad (1.39)$$

where a is the lift-curve gradient and α is the AOA. The product $a\alpha$ is called the *induced lift* (i.e. lift induced by AOA).

Conceptually, a two-dimensional (2D) wing is the cross-section of a wing with infinite span. The local flow around any section will show an identical pattern of streamlines,

Figure 1.7 Wing Force/Moment System.

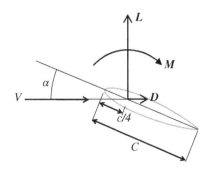

with no variation across the span. The 2D lift-curve gradient at low speed is crudely approximated as:

$$a_{2D} = 2\pi + 4.9\left(\frac{t}{c}\right) \tag{1.40}$$

where M is the flight Mach number and t/c is the thickness/chord ratio (which is often assumed to be 0.1 in the absence of real data). The 'ideal' lift-curve gradient for a thin wing is 2π.

A real wing has finite span and the flow over upper and lower surfaces has to mix in order to equalise the pressure at the wing tips. This is achieved by flow under the wing being drawn around the tips by the lower pressure above the wing. For the outer wing, the flow direction turns outwards over the lower surface and inwards over the upper surface. The 2D model is no longer usable because the flow has a three-dimensional (3D) pattern. The 3D lift-curve gradient for a rectangular wing [which was introduced as a in Equation 1.39] can be approximated using DATCOM[1]:

$$a = \frac{2\pi A}{2 + \sqrt{4 + (A/\kappa)^2\left(1 - M_N^2\right)}} \tag{1.41}$$

where M_N is the flight Mach number, A is the aspect ratio, and $\kappa = a_{2D}/2\pi$. The variation of this quantity is shown in Figure 1.8 for a wing with thickness/chord ratio of 0.1. It is worth noting that:

$$a \longrightarrow \frac{a_{2D}}{\sqrt{1 - M_N^2}} \qquad \text{as} \qquad A \longrightarrow \infty$$

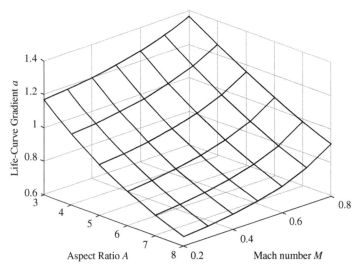

Figure 1.8 Variation of Lift-Curve Gradient.

---

1 Source: US Airforce Stability and Control Data Compendium (DATCOM).

The drag coefficient is written as the sum of profile drag, induced drag, and wave drag coefficients:

$$C_D = C_{D0} + C_{Di} + C_{Dw} \tag{1.42}$$

The main drag components are summarised as follows:

Profile drag is caused by air forcing its way around the outer geometry of the aircraft. In the absence of real data, a starting assumption might be $C_{D0} = 0.02$.

- Induced drag is caused by deflecting the airflow in order to generate lift. It is estimated as:

$$C_{Di} = \frac{C_L^2}{\pi A e} \tag{1.43}$$

where C_L is the lift coefficient, A is the wing aspect ratio, and e is the wing efficiency. Initial calculations might assume $\pi e \approx 2.5$ or thereabouts.

- Wave drag is caused by the shock waves as the Mach number M_N exceeds the critical Mach number M_{crit} (which is the maximum speed at which the flow field across the entire wing is subsonic). Lock's approximation is as follows:

$$C_{Dw} = 20 \left(M_N - M_{crit} \right)^4 \tag{1.44}$$

A rough estimate for M_{crit} is given by:

$$M_{crit} = k_T - \left(\frac{t}{c} \right) - 0.1 C_L - 0.1078 \tag{1.45}$$

where t/c is the thickness/chord ratio, C_L is the lift coefficient and the factor k_T is assigned a value of 0.87 (in the absence of specific data). This is discussed in Chapter 3.

1.5.3 Wing/Tail Combination

Figure 1.9 shows an aircraft in straight-and-level flight for a given velocity V and AOA α. Lift vectors are applied on the wing and tail. The aircraft CG is shown as a circle with alternate black and white quadrants. Engine thrust will contribute a vertical force and a pitching

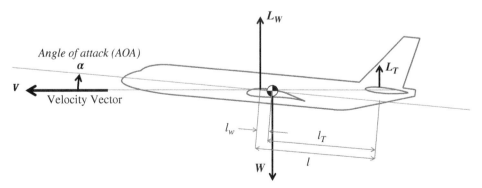

Figure 1.9 Aircraft in Straight-and-Level Flight.

moment to the aircraft but, for simplicity, it is convenient to neglect it (as textbooks usually do).

This aerodynamic force/moment system is evaluated with respect to the CG as follows:

$$L = L_W + L_T \qquad M = l_W L_W - l_T L_T \tag{1.46}$$

where subscripts W and T denote wing and tail, respectively. This can be written in coefficient form:

$$L = Q(S_W C_{LW} + S_T C_{LT}) \qquad M = Q(S_W l_W C_{LW} - S_T l_T C_{LT}) \tag{1.47}$$

where $Q = \rho V^2/2$ is the dynamic pressure, S_W is the wing planform area, and S_T is the tail planform area. The distances l_W and l_T define the respective positions of the wing and tail quarter-chord positions relative to the CG.

The lift coefficients are C_{LW} for the wing and C_{LT} for the tail, where:

$$C_{LW} = a_W \alpha_W \qquad C_{LT} = a_T \alpha_T \tag{1.48}$$

where a_W and a_T are the 3D lift-curve gradients for the wing and tail, respectively, and α_W and α_T are the corresponding angles of attack. In this simple aerodynamic model, the wing plane is aligned with the aircraft datum, such that:

$$\alpha_W = \alpha \tag{1.49}$$

As shown in Figure 1.10, airflow is deflected downwards behind the trailing edge (known as 'downwash') and, accordingly, the tail AOA is approximated as:

$$\alpha_T = \alpha - \varepsilon + \delta_T \tag{1.50}$$

where α is the aircraft AOA, ε is the downwash angle at the tail, and δ_T is the tail rotation angle (assuming that this aircraft has an all-moving tail). There is a time delay $t' = l_T/V$ for the air to flow from the wing to the tail, such that:

$$\alpha_T(t) = \alpha(t) - \varepsilon(t - t') + \delta_T(t)$$

Using a power series expansion for $\varepsilon(t - \tau)$, this becomes:

$$\alpha_T(t) = (1 - \varepsilon_a)\alpha(t) + t'\varepsilon_a \dot{\alpha}(t) + \delta_T(t) \tag{1.51}$$

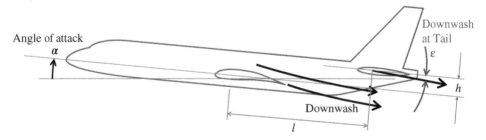

Figure 1.10 Effect of Downwash on Horizontal Tail.

The variation of downwash with AOA can be written as $\varepsilon = \varepsilon_\alpha \alpha$, where ε_α is the downwash gradient given by DATCOM:

$$\varepsilon_\alpha = \frac{d\varepsilon}{d\alpha} = 4.44 \, (K_A K_T)^{1.19} \tag{1.52}$$

where

$$K_A = \frac{1}{A_W} - \frac{1}{1 + A_W^{1.7}} \qquad K_T = \frac{1 - |h/b|}{\sqrt[3]{2l/b}}$$

The aspect ratio is A_W, the longitudinal separation of wing and tail is l and the vertical separation (height) is h. Note that l and h are measured relative to the wing chord plane, which is aligned with the aircraft datum in this simplified explanation. Note that Equation 1.51 is applicable to a wing with a rectangular planform. The general case is presented in Chapter 4.

The tail lift coefficient is now expressed as:

$$C_{LT} = a_T' \alpha + a_T (t' \varepsilon_\alpha \dot{\alpha} + \delta_T) \tag{1.53}$$

where the modified lift-curve gradient is $a_T' = a_T(1 - \varepsilon_\alpha)$.

1.5.4 Lift Distribution

Introductory texts do not usually consider the lift distribution over a wing because they deal with consolidated forces and moments. However, this provides a conceptually simple (albeit approximate) method for calculating rolling moments and, in Section 1.5.5, calculating the aerodynamics of flight controls. The general case is presented in Chapter 3 and a cut-down version is given here for a simple flat wing with a rectangular planform. So, the local lift coefficients are given by:

$$C_l(\eta) = C_L \, \partial(\eta) \tag{1.54}$$

where $\eta = y/(b/2)$ is the a nondimensional length parameter, C_L is the 3D lift coefficient, and $\partial(\eta)$ is the lift distribution function:

$$\partial(\eta) = \lambda + (1 - \lambda)\varepsilon(\eta) \tag{1.55}$$

where $e(\eta)$ is an elliptical distribution:

$$e(\eta) = \frac{4}{\pi} \sqrt{1 - \eta^2} \tag{1.56}$$

and the coefficient λ is a function of $f = (2\pi A/a_{2D})/14$, as shown in Figure 1.11:

$$\lambda \approx -0.1535 f^3 - 0.3121 f^2 + 1.4647 f \tag{1.57}$$

With reference to Equation 1.39, the local lift coefficient is determined as:

$$C_l(\eta) = \partial(\eta) a \alpha \tag{1.58}$$

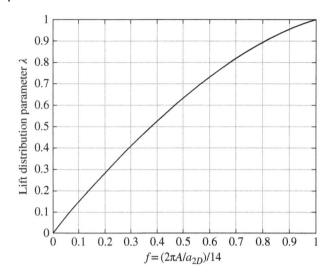

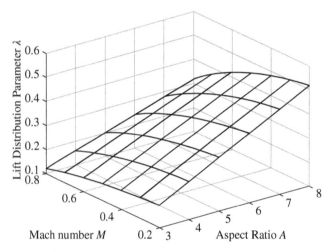

Figure 1.11 Variation of the Lift Distribution Coefficient.

This is valid for symmetric flight. In other condition, it is determined as:

$$C_l(\eta) = \partial(\eta)a\alpha_l(\eta) \tag{1.59}$$

where $\alpha_l(\eta)$ is the local AOA.

As explained in Chapter 3 and as indicated by Figure 1.12, the local AOA can be calculated as:

$$\alpha_l(\eta) = \alpha_s + \eta p' \tag{1.60}$$

where α_s is the symmetric AOA at the wing, $p' = bp/2V$ is the nondimensional roll rate, and η is the nondimensional length scale η. The local lift coefficient becomes:

$$C_l(\eta) = \partial(\eta)a(\alpha_s + \eta p') \tag{1.61}$$

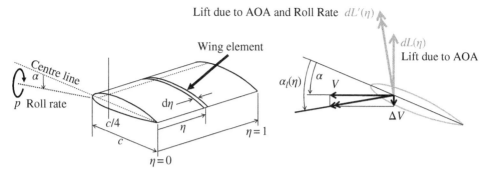

Figure 1.12   Incremental Lift on a Wing Element.

The total lift and total rolling moment on a wing is obtained as follows:

$$L = QS \int_0^1 C_l(\eta)d\eta \qquad K = -QS\frac{b}{2} \int_0^1 C_l(\eta)\eta\, d\eta \tag{1.62}$$

Lift is calculated for a symmetric distribution and rolling moment is calculated for an antisymmetric distribution (noting that a minus sign is needed because the right wing induces a left-handed moment about the velocity vector). Thus, the associated coefficients can be written as:

$$C_L = a(\alpha I_0(1) + p'I_1(1)) \qquad C_K = -a(\alpha I_1(1) + p'I_2(1)) \tag{1.63}$$

where

$$I_N(\eta_1) = \int_0^{\eta_1} \partial(\eta)\eta^N\, d\eta \tag{1.64}$$

Equivalent force vectors can be drawn on the starboard wing span, showing the lines of action for the total lift force and the total rolling moment. These are positioned at nondimensional distances η_L and η_K from the centre line, as defined by:

$$\eta_L = \frac{I_1(1)}{I_0(1)} \qquad \eta_K = \frac{I_2(1)}{I_1(1)} \tag{1.65}$$

Finally, recalling Equation 1.56:

$$\partial(\eta) = \lambda + (1-\lambda)e(\eta)$$

Therefore, the integral Δ_N is evaluated from 0 to $\eta_1 \leq 1$ as follows:

$$I_N(\eta_1) = \lambda C_N(\eta_1) + (1-\lambda)E_N(\eta_1) \tag{1.66}$$

where

$$C_N(\eta_1) = \int_0^{\eta_1} \eta^N\, d\eta \qquad E_N(\eta_1) = \int_0^{\eta_1} e(\eta)\eta^N\, d\eta \tag{1.67}$$

The distribution functions and their integrals are shown in Figure 1.13 for $N = 0, 1, 2$. For reference, $C_0(1) = 1$, $C_1(1) = 1/2$, $C_0(1) = 1/3$, $E_0(1) = 1$, $E_1(1) = 4/3\pi$ and $E_0(1) = 1/3$.

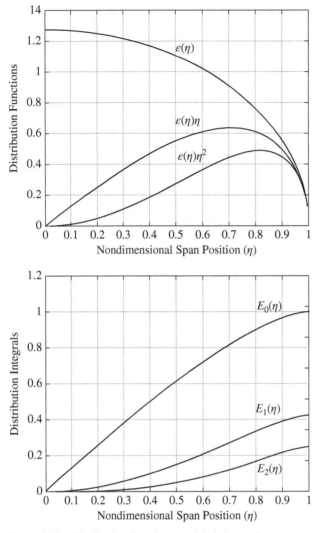

Figure 1.13   Distribution Functions and their Integrals.

1.5.5 Adding Flight Controls

Much has been written in textbooks on control surfaces and the body of empirical aerodynamics is considerable, covering incremental forces and moments together with correction factors for every applicable facet of real flow physics. Interested readers are referred to Raymer (1999) and Pamadi (2015), for example. This chapter will opt for a much simpler approach.

Flight controls modify the aerodynamics of a lifting surface. The discussion here is limited to plain flaps on the trailing edge, as drawn in Figure 1.14. These can be full-span or part-span, as seen in Figure 1.15. Symmetric deployment generates lift and antisymmetric deployment generates a rolling moment. Relevant parameters in the calculation of flap

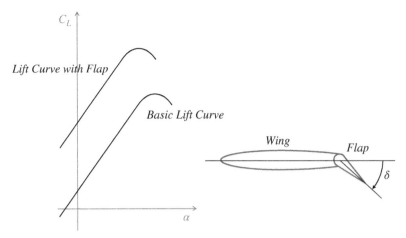

Figure 1.14 Changes in Lift Coefficient due to Flap Deflection.

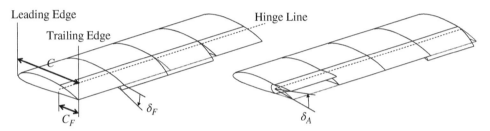

Figure 1.15 Trailing-Edge Control Surfaces.

effects are the chord length c (measured from leading to trailing edge) and flap chord length c_F (measured from hinge line to trailing edge). The flap-chord ratio is c_F/c.

The overall drag increment due to a control deflection δ is estimated in DATCOM as:

$$\Delta C_D = 1.7 \left(\frac{c_F}{c}\right)^{1.38} \left(\frac{S_F}{S}\right) \sin^2 \delta \tag{1.68}$$

where c_F/c is the flap-wing chord ratio and S_F/S is flap-wing area ratio. The value of S_F is the area calculated across the flap span and S is the area calculated across the wing span. The symbol δ would be replaced by δ_A for an aileron deflection, δ_F for a conventional wing flap deflection, or δ_E for an elevator deflection on the tail.

The aerodynamic effect of a control deflection can be construed as a change in the local AOA:

$$\Delta \alpha_l(\eta) = \alpha_\delta \delta \tag{1.69}$$

where α_δ is the flap effectiveness (shown in Figure 1.16), as calculated using the formula:

$$\alpha_\delta = 1 - \frac{\varphi - \sin \varphi}{\pi} \qquad \text{where} \qquad \cos \varphi = 2\frac{c_F}{c} - 1 \tag{1.70}$$

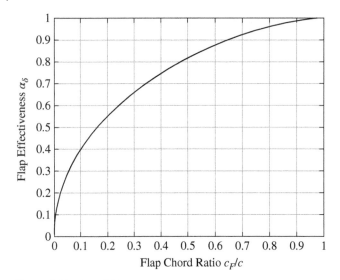

Figure 1.16 Flap Effectiveness.

Recalling Equation 1.58, the incremental change in the local lift coefficient is calculated as:

$$\Delta C_l(\eta) = \partial(\eta) a \alpha_\delta \delta_F \tag{1.71}$$

Recalling Equation 1.65, the resultant changes in lift force and rolling moment over the wing span $\eta_I \le \eta \le \eta_O$ (between the inboard and outboard edges of the control surface) are calculated as:

$$\Delta C_L = (I_0(\eta_O) - I_0(\eta_I))\alpha_\delta \delta \qquad \Delta C_K = -(I_1(\eta_O) - I_1(\eta_I))\alpha_\delta \delta \tag{1.72}$$

where the integrals I_0 and I_1 are defined in Equation 1.64.

The control deflection also induces a local pitching moment ΔC_m as follows:

$$\Delta C_m(\eta) = -K_m \Delta C_l(\eta) \tag{1.73}$$

where

$$K_m = 0.3077\left(0.805 - \frac{c_F}{c}\right)$$

This integrates to give the total change in pitching moment as follows:

$$C_M = -K_m \alpha_\delta \delta I_0 \tag{1.74}$$

1.6 Airspeed Definitions

Two definitions of airspeed are in common usage, derived from the true airspeed V. These are summarised below as the final component of the mathematical definition of this simple flight model.

Equivalent airspeed (V_{EAS}) at sea level corresponds with the dynamic pressure at a given altitude.

$$V_{EAS} = V \sqrt{\frac{\rho}{\rho_0}}$$

where the subscript 0 denotes Sea Level. Using relative density, as defined in Equation 1.16:

$$V_{EAS} = V \sqrt{\sigma} \tag{1.75}$$

Calibrated airspeed (V_{CAS}) at sea level corresponds with the impact pressure at a given altitude.

$$V_{CAS} = \sqrt{\frac{2\gamma}{\gamma - 1} R T_0 \left(\frac{Q_c}{P_0} + 1 \right)^{\frac{\gamma-1}{2\gamma}}} \tag{1.76}$$

where Q_c is the so-called impact pressure:

$$Q_c = P' - P \tag{1.77}$$

This is the difference between the stagnation pressure P' and the static pressure P. These two quantities are interrelated with the adiabatic gas constant γ and the Mach number M_N defined in Equation 1.15:

$$\frac{P'}{P} = \left(1 + \frac{\gamma - 1}{2} M_N^2 \right)^{\frac{\gamma}{\gamma-1}} \tag{1.78}$$

where M_N is the flight Mach number. Note that stagnation temperature T' is related to static temperature T as follows:

$$\frac{T'}{T} = 1 + \frac{\gamma - 1}{2} M_N^2 \tag{1.79}$$

These thermodynamic relationships are developed in Chapter 5 (with slightly different notation).

1.7   Flight Model Architecture

The structure/interface definition of any computational model is a matter of profound importance in the transcription of the model into software (however that is constituted). The principal objective is to partition the calculations into functionally concise blocks and, in so doing, to establish a dataflow map that is reasonably sparse. This is intended to render a model representation that is clear and readable (i.e. not complicated and not cluttered). One possible solution is shown in Figure 1.17, with blocks that correspond with the preceding Sections 1.2–1.5. The visible parameters are those that are transferred from one block to another: all others are contained within the relevant functional block as part of the required calculations.

In this particular model, the Motion and Propulsion blocks contain dynamics, with called state variables governed by differential equations [which are called state equations]. The universal principle is that the outputs from a functional block are fully determined by the current states and inputs. This is depicted in Figure 1.18.

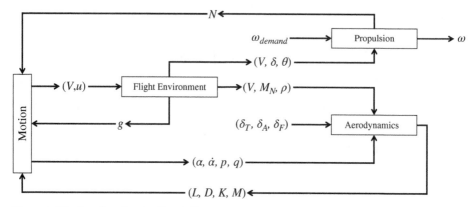

Figure 1.17 Dataflow for the Simple Flight Model.

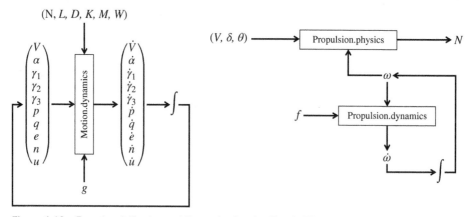

Figure 1.18 Functional Physics and Dynamics for the Simple Flight Model.

The Motion block receives inputs from Aerodynamics, Propulsion, and Flight Environment. These inputs are combined with the state vector, which generically is written as $\boldsymbol{x}$, in order to generate the rate of change $\dot{\boldsymbol{x}}$. In this case, Motion is comprised only of a 'dynamics' subblock and its outputs are the state variables (α, p, q) and the state derivative $\dot{\alpha}$.

The Propulsion block contains a performance calculation, which outputs nett thrust N, plus a state equation for engine speed. This partitions 'physics' and 'dynamics' in a way that many modelling textbooks and toolsets draw the distinction between 'output equations' and 'state equations', respectively. Note that the inputs (V, δ, θ) are taken from the Flight Environment and the control input f is received from outside the model.

There are many ways in which a model like this can be organised and the intent here is NOT to present a methodology, only to indicate the sort of considerations that form part of the underlying thought process. A functional block diagram is an appropriate representation for the Simple Flight Model because it shows clearly where calculations are performed and where parameters are transferred. A summary of 'visible' data is given in Table 1.2. In a full development project, this would be broken down into block-to-block transfers and

Table 1.2 Interface Parameters in the Simple Flight Model.

Symbol	Definition	Units
α	Angle of attack (AOA) [cf. Figure 1.1]	radian
δ	Relative pressure	Not applicable
θ	Relative temperature	Not applicable
ρ	Air density	$kg\,m^{-3}$
g	Gravitational acceleration	$m\,s^{-2}$
m	Aircraft mass	kg
p	Roll rate (about velocity vector) [cf. Figure 1.4]	$radian\,s^{-1}$
q	Pitch rate [cf. Figure 1.4]	$radian\,s^{-1}$
u	Altitude measured up from earth surface	m
D	Aerodynamic drag	N
J_1	Aircraft moment of inertia in roll	$kg.m^2$
J_2	Aircraft moment of inertia in pitch	$kg.m^2$
L	Aerodynamic lift	N
K	Aerodynamic rolling moment	N.m
M	Aerodynamic pitching moment	N.m
M_N	Mach number	Not applicable
N	Nett thrust	N
V	True airspeed [cf. Figure 1.3]	$m\,s^{-1}$
W	Aircraft weight	N

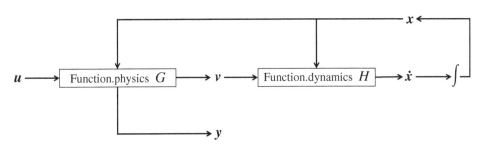

Figure 1.19 Physics and Dynamics of a Generic Functional Block.

input-to-output transformations within each block. The end-result would be a comprehensive statement of content for the model, as a formally declared product model.

A final comment in this Chapter concerns generic computational structure. Figure 1.19 shows one way of defining the contents of a functional block, with the partition of physics and dynamics that has been discussed already. Inputs are supplied as a vector $\boldsymbol{u}$, outputs are delivered as a vector $\boldsymbol{y}$, states are stored in a vector $\boldsymbol{x}$, and state derivatives are updated to a vector $\dot{x}$. The interface between physics and dynamics is via a vector v. The constitutive relationships for this scheme are as follows:

$$(v, y) = G(u, x) \tag{1.80}$$

$$\dot{x} = H(v, x) \tag{1.81}$$

where G and H are non-linear functions. Small perturbations about an equilibrium condition would be described in the general form:

$$\dot{x} = Ax + Bv \qquad y = Cx + Du \qquad v = Ex + Fu \tag{1.82}$$

This is rationalised as follows:

$$\dot{x} = A'x + B'u \qquad y = Cx + Du \tag{1.83}$$

where $A' = A + BE$ and $B' = BF$, which is the form that is most familiar to dynamicists and control engineers.

2

Equations of Motion

2.1 Introduction

2.1.1 The Problem with Equations of Motion

Everybody understands equations of motion ... or they should! At school, everybody learns Newtons' Laws of Motion and, at higher levels of education, these are generalised and applied in different contexts. It is not entirely clear how much of this subject is understood by students or engineers although, in fairness, most individuals do not need expert insight. However, flight dynamicists do need a better-than-average grasp of the applicable science and mathematics and they need to understand the various formulations and foundations for representing the motion of a vehicle through space.

One televised documentary include commentary from a physics teacher who 'explained' flight wholly in relation to Newton's Third Law, which is wrong. A prominent researcher proclaimed that 'nobody uses quaternions', which is also wrong. A lecturer once claimed that aerospace students do not need to study this subject because companies always employ physics graduates to do this kind of work. Wrong again! While experts will be acutely aware of mis-interpretations and confusions, others may not be. So, in the interest of educating the next generation of experts, this author believes that textbooks should present a complete explanation of Equations of Motion. This is because simulation has to predict aircraft motion and, to that end, it is important to know what formulation is being applied and what computational is being performed.

2.1.2 What Chapter 2 Includes

This chapter includes:

- Spatial Reference Model
  (covering reference frames for earth, aircraft, and flight path)
- Aircraft Dynamics
  (developing equations of motion in terms of force, moment, and velocity components)
- Aircraft Kinematics
  (developing equations of motion for position and orientation)

- Initialisation
  (establishing equilibrium flight conditions by balancing forces and moments)
- Linearisation
  (recasting the equations of motion for small perturbations about equilibrium)

2.1.3 What Chapter 2 Excludes

This chapter excludes:

- Nondimensional equations of motion
  (which could be readily derived but do not add heuristic value)
- Systematic treatment of numerical methods applied for initialisation and linearisation
  (which could be explained but better explained properly in a dedicated math textbook)

2.1.4 Overall Aim

Chapter 2 should provide a complete development of the equations of motion for aircraft dynamics and kinematics, suitable for implementation as part of an aircraft simulation. This includes a method for initialising aircraft flight parameters for a given flight condition and a method for producing linearised equations that are used in linear analysis (typically for flight control law design).

2.2 Spatial Reference Model

The physics of aircraft motion are based on Newton's Second Law and a set of kinematic equations that determine position and orientation in space. All of this resides within a spatial reference model that is comprised of reference frames, which are otherwise known as axis systems. This is summarised here as a set of tutorial notes in order to provide the context and the pre-requisites for developing the equations of motion for aircraft in flight. For more detail, refer to Volume 1.

2.2.1 Generic Reference Frames

A reference frame is an *xyz* axis system with basis vectors that are mutually orthogonal (i.e. at right angles to each other). Relative orientation is interpreted in Figure 2.1 as the relationship between two frames denoted by superscripts '0' and '1'. Notationally, it is convenient to define frames $\mathcal{F}^0$ and $\mathcal{F}^1$ by assembling basis vectors into matrices as follows:

$$\mathcal{F}^0 = \begin{bmatrix} x^0, y^0, z^0 \end{bmatrix} \qquad \mathcal{F}^1 = \begin{bmatrix} x^1, y^1, z^1 \end{bmatrix} \tag{2.1}$$

These frames are interrelated by a physical rotation from $\mathcal{F}^1$ to $\mathcal{F}^2$:

$$\mathcal{F}^1 = \mathcal{F}^0 R^{01} \tag{2.2}$$

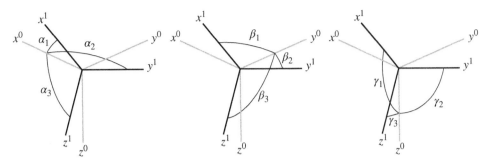

Figure 2.1 Relative Orientation.

The rotation matrix R^{01} is a direction cosine matrix:

$$R^{01} = \begin{pmatrix} \cos\alpha_1 & \cos\alpha_2 & \cos\alpha_3 \\ \cos\beta_1 & \cos\beta_2 & \cos\beta_3 \\ \cos\gamma_1 & \cos\gamma_2 & \cos\gamma_3 \end{pmatrix} \tag{2.3}$$

The columns contain the axis vectors of $\mathcal{F}^1$ are measured with respect to the axis vectors of $\mathcal{F}^0$. This matrix is orthogonal and, as such, its inverse and transpose are identical:

$$\left(R^{01}\right)^{-1} = \left(R^{01}\right)^{T} \equiv R^{10} \tag{2.4}$$

The inverse rotation is from $\mathcal{F}^1$ to $\mathcal{F}^0$ and, thus, $R^{01}R^{10} = I$, where I is the identity matrix. If $\mathcal{F}^0$ is not contained within another frame, then it is defined by:

$$x^0 = \begin{pmatrix} 1 \\ 0 \\ 0 \end{pmatrix} \quad y^0 = \begin{pmatrix} 0 \\ 1 \\ 0 \end{pmatrix} \quad z^0 = \begin{pmatrix} 0 \\ 0 \\ 1 \end{pmatrix} \quad \Longrightarrow \quad \mathcal{F}^0 = \begin{pmatrix} 1 & 0 & 0 \\ 0 & 1 & 0 \\ 0 & 0 & 1 \end{pmatrix} \tag{2.5}$$

Otherwise, its basis vectors are measured with respect to the basis vectors of a parent frame.

A vector v is always defined with respect to a reference frame, for instance:

$$v = \mathcal{F}^0 v^0 \tag{2.6}$$

It can be expressed with respect to another frame as follows:

$$v = \mathcal{F}^0 v^0 = \mathcal{F}^0 R^{01} R^{10} v^0 = \mathcal{F}^1 v^1 \tag{2.7}$$

Thus, the distinction is drawn between *physical transformations* that define the orientation of the basis vectors:

$$\mathcal{F}^1 = \mathcal{F}^0 R^{01} \qquad \mathcal{F}^0 = \mathcal{F}^1 R^{10}$$

and *coordinate transformations* that resolve vector coordinates with respect to the basis vectors:

$$v^0 = R^{01}v^1 \qquad v^1 = R^{10}v^0$$

2.2.2 Rotating Reference Frames

A reference frame can rotate with respect to its parent although, ultimately, it can always be traced back to a non-rotating, non-accelerating frame. This is called an *inertial frame*.

If $\mathcal{F}^1$ is rotating with respect to an inertial frame $\mathcal{F}^i$, then a fixed vector in $\mathcal{F}^1$ is not a fixed vector in $\mathcal{F}^i$. The rate of change of a vector depends on the frame in which it is defined and the frame in which it is being measured.

Figure 2.2 shows these two frames, $\mathcal{F}^i$ (which does not rotate) and $\mathcal{F}^1$ (which rotates with respect to $\mathcal{F}^i$ at a rate defined by an angular velocity vector ω^1). This vector projects outwards from the plane of the image and so a 'positive' rotation is counter-clockwise.

Vector v^1 has components defined in $\mathcal{F}^1$. The change in this vector is Δv^1, measured over an infinitesimal time interval Δt. When that change is added to the frame rotation, the resulting change measured in the inertial frame is $\Delta v^{1,i}$ (indicated explicitly by an extra superscript 'i'). The difference between these measurements is due to the rotation of $\mathcal{F}^1$ through an angle of $\omega^1 \Delta t$.

Mathematically, this is summarised as:

$$\Delta v^{1,i} = \Delta v^1 + \left(\omega^1 \Delta t\right) \times v^1 \tag{2.8}$$

Equivalently:

$$\frac{\Delta v^{1,i}}{\Delta t} = \frac{\Delta v^1}{\Delta t} + \omega^1 \times v^1$$

When $\Delta t \longrightarrow 0$, the vector derivative becomes:

$$\dot{v}^{1,i} = \dot{v}^1 + \omega^1 \times v^1 \tag{2.9}$$

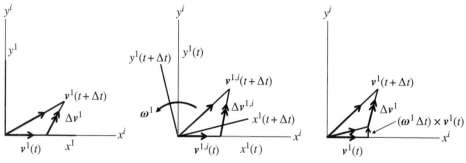

Figure 2.2 Vector Differentiation.

This is written as a matrix equation, as follows:

$$\dot{v}^{1,i} = \dot{v}^1 + \Omega_1 v^1 \tag{2.10}$$

where Ω_1 is called the cross-product matrix, which satisfies the equation $\Omega_1 v^1 = \omega^1 \times v^1$.

Now, it is possible to have another rotating frame, such that a new frame $\mathcal{F}^2$ rotates within $\mathcal{F}^1$, which in turn rotates within $\mathcal{F}^i$. Note that $\mathcal{F}^2$ sees $\mathcal{F}^1$ as non-rotating and, thus the local vector derivative is:

$$\dot{v}^{2,1} = \dot{v}^2 + \Omega_2 v^2 \tag{2.11}$$

Relevant transformations are, as follows:

$$v^2 = R^{21} v^1 \qquad v^1 = R^{12} v^2 \qquad \dot{v}^{2,1} = R^{21} \dot{v}^1 \tag{2.12}$$

Equations 2.17, 2.18 and 2.19 are combined, as follows:

$$\dot{v}^{2,i} = R^{21} \dot{v}^{1,i}$$
$$\dot{v}^{2,i} = R^{21} \left(\dot{v}^1 + \Omega_1 v^1 \right)$$
$$\dot{v}^{2,i} = R^{21} \left(\dot{v}^1 + \Omega_1 R^{12} v^2 \right)$$
$$\dot{v}^{2,i} = \dot{v}^{2,1} + R^{21} \Omega_1 R^{12} v^2$$

Therefore:

$$\dot{v}^{2,i} = \dot{v}^2 + \left(\Omega_2 + R^{21} \Omega_1 R^{12} \right) v^2 \tag{2.13}$$

This can be extended to any number of frames although this chapter only needs to deal with two.

2.2.3 Elementary Rotations

When a rotations are applied about individual axes, they are called elementary rotations. These were developed in Volume 1 as *coordinate transformations*:

$$R_x(\delta) = \begin{pmatrix} 1 & 0 & 0 \\ 0 & \cos\delta & -\sin\delta \\ 0 & \sin\delta & \cos\delta \end{pmatrix} \quad R_y(\delta) = \begin{pmatrix} \cos\delta & 0 & \sin\delta \\ 0 & 1 & 0 \\ -\sin\delta & 0 & \cos\delta \end{pmatrix} \quad R_z(\delta) = \begin{pmatrix} \cos\delta & \sin\delta & 0 \\ -\sin\delta & \cos\delta & 0 \\ 0 & 0 & 1 \end{pmatrix} \tag{2.14}$$

In Volume 2, they are constructed as *physical transformations*:

$$R^X(\delta) = \begin{pmatrix} 1 & 0 & 0 \\ 0 & \cos\delta & \sin\delta \\ 0 & -\sin\delta & \cos\delta \end{pmatrix} \quad R^Y(\delta) = \begin{pmatrix} \cos\delta & 0 & -\sin\delta \\ 0 & 1 & 0 \\ \sin\delta & 0 & \cos\delta \end{pmatrix} \quad R^Z(\delta) = \begin{pmatrix} \cos\delta & -\sin\delta & 0 \\ \sin\delta & \cos\delta & 0 \\ 0 & 0 & 1 \end{pmatrix} \tag{2.15}$$

Comparing Equations 2.8 and 2.9, a coordinate transformation is the transpose (and inverse) of a physical transformation.

$$R_x(\delta) = R^X(-\delta) \qquad R_y(\delta) = R^Y(-\delta) \qquad R_z(\delta) = R^Z(-\delta) \tag{2.16}$$

Any rotation in space can be decomposed into a sequence of elementary rotations

2.2.4 Reference Frames for Position and Orientation

Figure 2.3 shows the three reference frames that are required in order to represent aircraft position and orientation with respect to the Earth. It is noted that the WGS84[1] Earth rotation rate is stated as $\omega_E = 7.292115 \times 10^{-5}$ rad s^{-1}, which is equivalent to just under 361° per day (computed as 360.9856050255709).

Frame $\mathcal{F}^0$ is aligned with the Equator and the Greenwich meridian. It rotates with the Earth and is designated as Earth Centred Earth Fixed (ECEF).

Frame $\mathcal{F}^1$ defines the local ground plane, with its origin directly beneath the aircraft and with its orientation determined by longitude λ and latitude μ. It is designated as North-East-Down (NED) because of the alignemnt of its basis vectors. The transformation from $\mathcal{F}^0$ to $\mathcal{F}^1$ is specified by the sequence of elementary rotations in Table 2.1:

$$R^{01} = R^Z(\lambda)R^Y(-\mu)R^Y\left(-\frac{\pi}{2}\right) \tag{2.17}$$

Figure 2.3 Reference Frames.

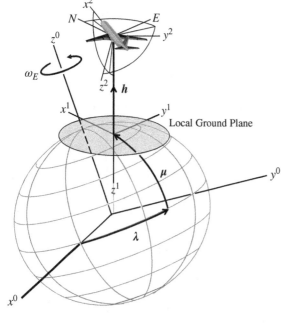

Local Ground Plane

Table 2.1 ECEF-to-NED Rotations.

	Elementary Rotation	Angular Range	Direction
1	Longitude	$0 \leq \lambda < 2\pi$	Clockwise about the **z** axis of ECEF
2	Geodetic latitude	$-\pi/2 \leq \mu \leq \pi/2$	Clockwise about the resulting **y** axis
3	Alignment with NED	$-\pi/2$	Clockwise about the resulting **y** axis

Note that, after two rotations, the frame is aligned such that x points up, y points east and z points north. So, a re-orientation is needed for *xyz* axes to be correctly aligned with North-East-Down.

Frame $\mathcal{F}^2$ defines the Aircraft Datum Frame, as shown in Figure 2.4. The origin is located at the aircraft datum point, which usually coincides with a conveniently placed structural frame. Elementary rotations are used to define aircraft orientation, as specified in Table 2.2:

$$R^{12} = R^Z(\psi)R^Y(\theta)R^X(\varphi) \tag{2.18}$$

Figure 2.4 Aircraft Orientation.

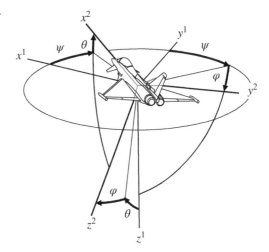

Table 2.2 NED-to-Aircraft Rotations.

	Elementary Rotation	Angular Range	Direction
1	Azimuth angle	$0 \leq \psi < 2\pi$	Clockwise about the z axis of NED
2	Pitch angle	$-\pi/2 \leq \theta \leq \pi/2$	Clockwise about the resulting y axis
3	Roll angle	$-\pi < \varphi \leq \pi$	Clockwise about the resulting x axis

The rotation angles are *Euler Angles* when applied in this sequence (namely azimuth followed by pitch, followed by roll).

The physical transformation $\mathcal{F}^1$ to $\mathcal{F}^2$ is expanded as:

$$R^{12} = \begin{pmatrix} \cos\theta\cos\psi & \sin\varphi\sin\theta\cos\psi - \cos\varphi\sin\psi & \cos\varphi\sin\theta\cos\psi + \sin\varphi\sin\psi \\ \cos\theta\sin\psi & \sin\varphi\sin\theta\sin\psi + \cos\varphi\cos\psi & \cos\varphi\sin\theta\sin\psi - \sin\varphi\cos\psi \\ -\sin\theta & \sin\varphi\cos\theta & \cos\varphi\cos\theta \end{pmatrix}$$

$$(2.19)$$

Azimuth, pitch and roll angles can be determined from elements of matrix R^{12}, as follows:

$$\sin\theta = -R^{12}(3,1) \qquad \tan\varphi = \frac{R^{12}(3,2)}{R^{12}(3,3)} \qquad \tan\varphi = \frac{R^{12}(3,2)}{R^{12}(3,3)} \qquad (2.20)$$

2.2.5 Reference Frame for Flight Path

Figure 2.5 shows the NED frame ($\mathcal{F}^1$) and a new frame for the aircraft flight path, as defined by the sequence of elementary rotations in Table 2.3. Frame $\mathcal{F}^3$ has its x-axis aligned with the velocity vector and z-axis aligned with the aerodynamic lift vector (in the opposite direction). The transformation from $\mathcal{F}^1$ to $\mathcal{F}^3$ is:

$$R^{13} = R^Z(\gamma_3)R^Y(\gamma_2)R^X(\gamma_1) \qquad (2.21)$$

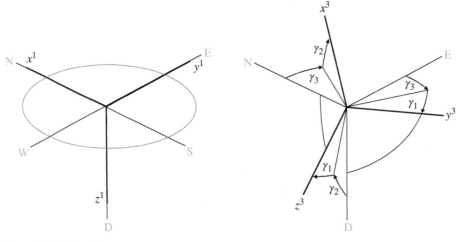

Figure 2.5 Flight Path Axes.

Table 2.3 Flight Path Rotations.

	Elementary Rotation	Angular Range	Direction
1	Track angle	$0 \leq \gamma_3 < 2\pi$	Clockwise about the z axis of NED
2	Climb/dive angle	$-\pi/2 \leq \gamma_2 \leq \pi/2$	Clockwise about the resulting y axis
3	Bank angle	$-\pi < \gamma_1 \leq \pi$	Clockwise about the resulting x axis

Applying matrix algebra, this transformation is expanded as:

$$R^{13} = \begin{pmatrix} \cos\gamma_2\cos\gamma_3 & \sin\gamma_1\sin\gamma_2\cos\gamma_3 - \cos\gamma_1\sin\gamma_3 & \cos\gamma_1\sin\gamma_2\cos\gamma_3 + \sin\gamma_1\sin\gamma_3 \\ \cos\gamma_2\sin\gamma_3 & \sin\gamma_1\sin\gamma_2\sin\gamma_3 + \cos\gamma_1\cos\gamma_3 & \cos\gamma_1\sin\gamma_2\sin\gamma_3 - \sin\gamma_1\cos\gamma_3 \\ -\sin\gamma_2 & \sin\gamma_1\cos\gamma_2 & \cos\gamma_1\cos\gamma_2 \end{pmatrix}$$

$$(2.22)$$

2.2.6 Airspeed and Airstream Direction

The velocity vector V gives the movement of an aircraft through the air. It combines airspeed (V), angle of attack or AOA (α) and angle of sideslip or AOS (β), as in Figure 2.6. These quantities are defined as:

$$\tan\alpha = \frac{w^1}{u^1} \qquad \sin\beta = \frac{v^1}{V^1} \qquad V = \sqrt{(u^1)^2 + (v^1)^2 + (w^1)^2} \qquad (2.23)$$

If the aircraft axes were rotated through the sequence in Table 2.4, they would align with the flight path axes. Thus, the transformation from $\mathcal{F}^2$ to $\mathcal{F}^3$ is, as follows:

$$R^{23} = R^Y(-\alpha)R^Z(\beta) \qquad (2.24)$$

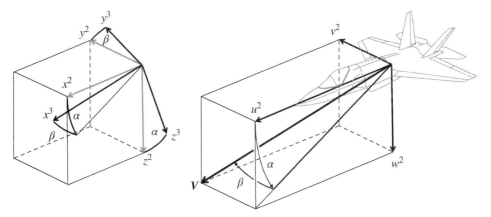

Figure 2.6 Airflow Parameters.

Table 2.4 Airstream Direction.

	Elementary Rotation	Angular Range	Direction
1	Angle of attack	$-\pi < \alpha \leq \pi$	Anti-clockwise about the y axis of ABC
2	Angle of sideslip	$-\pi/2 \leq \beta \leq \pi/2$	Clockwise about the resulting z axis

This is expanded to give:

$$R^{23} = \begin{pmatrix} \cos\alpha\cos\beta & -\cos\alpha\sin\beta & -\sin\alpha \\ \sin\beta & \cos\beta & 0 \\ \sin\alpha\cos\beta & -\sin\alpha\sin\beta & \cos\alpha \end{pmatrix} \tag{2.25}$$

2.3 Aircraft Dynamics

2.3.1 Mass Properties

The *mass properties* of an aircraft are mass (m), moments of inertia (J_x, J_y, J_z), products of inertia (J_{xy}, J_{yz}, J_{zx}) and the coordinates of the centre of gravity ($\bar{x}, \bar{y}, \bar{z}$). All of these quantities derive from summations performed on a mass distribution, i.e. an aircraft that has been broken down into infinitesimal elements of mass dm. Each element has coordinates (x, y, z) measured from the aircraft datum. A set of conventional definitions follow:

$$m = \int dm \qquad m\bar{x} = \int x\,dm \qquad m\bar{y} = \int y\,dm \qquad m\bar{z} = \int z\,dm \tag{2.26}$$

The kinetic energy associated with each mass element can be written as:

$$dE = \frac{1}{2}\mathbf{V}^T dm\,\mathbf{V} \tag{2.27}$$

where the velocity is determined by the angular velocity ω measured at the origin and by the position vector $\mathbf{p} = (x, y, z)^T$ of the mass element. Thus,

$$\mathbf{V} = \mathbf{p} \times \omega = P\omega \tag{2.28}$$

where the cross-product matrix is:

$$P = \begin{pmatrix} 0 & -z & y \\ z & 0 & -x \\ -y & x & 0 \end{pmatrix} \tag{2.29}$$

The kinetic energy becomes:

$$dE = \frac{1}{2}\omega^T dJ\,\omega \tag{2.30}$$

where the associated inertia matrix is given by:

$$dJ = P^T dm\, P = P^T P\, dm$$

$$dJ = \begin{pmatrix} y^2 + z^2 & -zx & -xy \\ -zx & z^2 + x^2 & -yz \\ -xy & -yz & x^2 + y^2 \end{pmatrix} dm \qquad (2.31)$$

The full inertia matrix is constructed as:

$$J = \begin{pmatrix} \int (y^2 + z^2)\,dm & -\int zx\,dm & -\int xy\,dm \\ -\int zx\,dm & \int (z^2 + x^2)\,dm & -\int yz\,dm \\ -\int xy\,dm & -\int yz\,dm & \int (x^2 + y^2)\,dm \end{pmatrix} \qquad (2.32)$$

Equivalently, this is written, as follows:

$$J = \begin{pmatrix} J_x & -J_{xy} & -J_{zx} \\ -J_{xy} & J_y & -J_{yz} \\ -J_{zx} & -J_{yz} & J_z \end{pmatrix} \qquad (2.33)$$

Mass properties will be discussed further in Chapter 6.

2.3.2 Flight Parameters

Figure 2.7 shows the aircraft axis system together with the nomenclature for flight dynamics. Axes are fixed within the aircraft and the *xyz* directions are designated as *forward*,

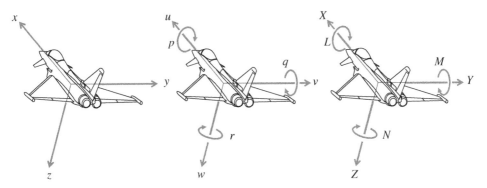

Figure 2.7 Aircraft Flight Parameters.

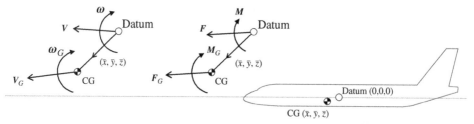

Figure 2.8 Defining Flight Parameters at the Aircraft Datum and the Aircraft CG.

lateral, and *vertical,* respectively. Flight parameters are the components of linear velocity u, v, w, angular velocity p, q, r, force X, Y, Z, and moment L, M, N:

$$F = \begin{pmatrix} X \\ Y \\ Z \end{pmatrix} \qquad M = \begin{pmatrix} L \\ M \\ N \end{pmatrix} \qquad V = \begin{pmatrix} u \\ v \\ w \end{pmatrix} \qquad \omega = \begin{pmatrix} p \\ q \\ r \end{pmatrix} \qquad (2.34)$$

In addition, linear momentum μ and angular momentum h are defined as follows:

$$\begin{pmatrix} \mu \\ h \end{pmatrix} = \begin{pmatrix} mI & 0 \\ 0 & J \end{pmatrix} \begin{pmatrix} V \\ \omega \end{pmatrix} \qquad (2.35)$$

where m is the (scalar) aircraft mass, I is the identity matrix, and J is the aircraft inertia matrix.

Figure 2.8 shows flight parameters referred to both the aircraft datum at $(0, 0, 0)$ and the CG at $(\bar{x}, \bar{y}, \bar{z})$. In computatonal models, force, moment, linear velocity, and angular velocity are all calculated at the aircraft datum. These quantities are resolved at the CG, as follows:

$$\begin{pmatrix} F \\ M \end{pmatrix}_G = \begin{pmatrix} I & 0 \\ G & I \end{pmatrix} \begin{pmatrix} F \\ M \end{pmatrix} \qquad \begin{pmatrix} V \\ \omega \end{pmatrix}_G = \begin{pmatrix} I & G \\ 0 & I \end{pmatrix} \begin{pmatrix} V \\ \omega \end{pmatrix} \qquad (2.36)$$

The inverse relationships are:

$$\begin{pmatrix} F \\ M \end{pmatrix} = \begin{pmatrix} I & 0 \\ -G & I \end{pmatrix} \begin{pmatrix} F \\ M \end{pmatrix}_G \qquad \begin{pmatrix} V \\ \omega \end{pmatrix} = \begin{pmatrix} I & -G \\ 0 & I \end{pmatrix} \begin{pmatrix} V \\ \omega \end{pmatrix}_G \qquad (2.37)$$

Matrix G is the cross-product matrix:

$$G = \begin{pmatrix} 0 & -\bar{z} & \bar{y} \\ \bar{z} & 0 & -\bar{x} \\ -\bar{y} & \bar{x} & 0 \end{pmatrix} \qquad (2.38)$$

2.3.3 Dynamic Equations of Motion

In its classical form, Newton's second law of motion for a physical object gives the relationship between force F and momentum μ in a non-rotating frame. It also gives the relationship between moment **M** and angular momentum **h**. The combined equations of motion are:

$$\begin{pmatrix} F \\ M \end{pmatrix}_G = \begin{pmatrix} \dot{\mu} \\ \dot{h} \end{pmatrix}_G \tag{2.39}$$

Recalling the discussion in Section 2.1.2, vector differentiation leads to generalised equations:

$$\begin{pmatrix} F \\ M \end{pmatrix}_G = \begin{pmatrix} \dot{\mu} \\ \dot{h} \end{pmatrix}_G + \begin{pmatrix} \Omega & 0 \\ 0 & \Omega \end{pmatrix} \begin{pmatrix} \mu \\ h \end{pmatrix}_G \tag{2.40}$$

where

$$\Omega = \Omega_2 + R^{20}\Omega_0 R^{02}$$

and where

$$\Omega_2 = \begin{pmatrix} 0 & -r & q \\ r & 0 & -p \\ -q & p & 0 \end{pmatrix} \qquad \Omega_0 = \begin{pmatrix} 0 & -\omega_E & 0 \\ \omega_E & 0 & 0 \\ 0 & 0 & 0 \end{pmatrix}$$

Momentum is defined in Equation 2.35 and is applied at the aircraft CG:

$$\begin{pmatrix} \mu \\ h \end{pmatrix}_G = \begin{pmatrix} mI & 0 \\ 0 & J \end{pmatrix} \begin{pmatrix} V \\ \omega \end{pmatrix}_G \tag{2.41}$$

Its derivative is as follows:

$$\begin{pmatrix} \dot{\mu} \\ \dot{h} \end{pmatrix}_G = \begin{pmatrix} \dot{m} & 0 \\ 0 & \dot{J} \end{pmatrix} \begin{pmatrix} V \\ \omega \end{pmatrix}_G + \begin{pmatrix} m & 0 \\ 0 & J \end{pmatrix} \begin{pmatrix} \dot{V} \\ \dot{\omega} \end{pmatrix}_G$$

After substitution and re-arrangement, the equations of motion become:

$$\begin{pmatrix} F \\ M \end{pmatrix}_G = \begin{pmatrix} \dot{m} + \Omega m & 0 \\ 0 & \dot{J} + \Omega J \end{pmatrix} \begin{pmatrix} V \\ \omega \end{pmatrix}_G + \begin{pmatrix} m & 0 \\ 0 & J \end{pmatrix} \begin{pmatrix} \dot{V} \\ \dot{\omega} \end{pmatrix}_G \tag{2.42}$$

The mapping of vector quantities from aircraft datum to CG is given by Equation 2.37:

$$\begin{pmatrix} F \\ M \end{pmatrix}_G = \begin{pmatrix} I & 0 \\ G & I \end{pmatrix} \begin{pmatrix} F \\ M \end{pmatrix} \qquad \begin{pmatrix} V \\ \omega \end{pmatrix}_G = \begin{pmatrix} I & G \\ 0 & I \end{pmatrix} \begin{pmatrix} V \\ \omega \end{pmatrix}$$

Accelerations are derived as:

$$\begin{pmatrix} \dot{V} \\ \dot{\omega} \end{pmatrix}_G = \begin{pmatrix} I & G \\ 0 & I \end{pmatrix} \begin{pmatrix} \dot{V} \\ \dot{\omega} \end{pmatrix} + \begin{pmatrix} 0 & \dot{G} \\ 0 & 0 \end{pmatrix} \begin{pmatrix} V \\ \omega \end{pmatrix}$$

After substitution and re-arrangement, the equations of motion are constructed in their final form:

$$\begin{pmatrix} m & mG \\ 0 & J \end{pmatrix} \begin{pmatrix} \dot{V} \\ \dot{\omega} \end{pmatrix} = \begin{pmatrix} I & 0 \\ G & I \end{pmatrix} \begin{pmatrix} F \\ M \end{pmatrix} - \begin{pmatrix} \dot{m} + m\Omega & (\dot{m} + m\Omega)G + m\dot{G} \\ 0 & \dot{J} + \Omega J \end{pmatrix} \begin{pmatrix} V \\ \omega \end{pmatrix}$$

(2.43)

For clarity, this can be re-packaged as:

$$E_{dyn}\dot{x}_{dyn} = A_{dyn}x_{dyn} + B_{dyn}u_{dyn}$$

(2.44)

The state and input vectors are:

$$x_{dyn} = \begin{pmatrix} V \\ \omega \end{pmatrix} \qquad u_{dyn} = \begin{pmatrix} F \\ M \end{pmatrix}$$

(2.45)

The 'mass matrix' E, the state transition matrix A, and the input matrix B are defined as:

$$E_{dyn} = \begin{pmatrix} m & mG \\ 0 & J \end{pmatrix} \qquad A_{dyn} = -\begin{pmatrix} \dot{m} + \Omega m & (\dot{m} + \Omega m)G + m\dot{G} \\ 0 & \dot{J} + \Omega J \end{pmatrix} \qquad B_{dyn} = \begin{pmatrix} I & 0 \\ G & I \end{pmatrix}$$

(2.46)

This format will be very familiar to anyone who works in dynamics and control. However, please note that Equation 2.40 is nonlinear (i.e. its matrix coefficients are variable, not constant).

The dataflow associated with aircraft dynamics is summarised in Figure 2.9.

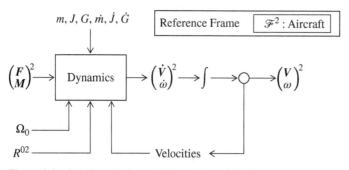

Figure 2.9 Dataflow for Dynamic Equations of Motion.

2.4 Aircraft Kinematics

2.4.1 Aircraft Position

The idealised shape of the Earth is defined by the WGS84 Ellipsoid, with parameters given in Table 2.5. From Figure 2.3, the aircraft datum point is located by longitude (λ), geodetic latitude (μ), and altitude (h). This point is located by ECEF coordinates:

$$x = (N + h)\cos\mu\cos\lambda$$
$$y = (N + h)\cos\mu\sin\lambda \qquad (2.47)$$
$$z = \left[N(1 - e^2) + h\right]\sin\mu$$

where N is the length of the *prime vertical* (which is the perpendicular line from the local ground plane to its point of intersection with the polar axis):

$$N = \frac{a}{\sqrt{1 - e^2\sin^2\mu}} \qquad (2.48)$$

Bowring (1976) provides an inverse calculation so that longitude, latitude, and altitude can be calculated from coordinates. Longitude is determined as:

$$\tan\lambda = \frac{y}{x} \qquad (2.49)$$

Latitude is initialised as μ_0:

$$\tan\mu_0 = \frac{z}{w} \qquad \text{where} \qquad w = \sqrt{x^2 + y^2} \qquad (2.50)$$

An iterative process calculates μ_n for $n = 0, 1, 2, \ldots$ until successive values of μ_n converge to within a predefined tolerance. The computational cycle is as follows:

$$\tan\theta_n = \left(\frac{c}{a}\right)\tan\mu_n \qquad \tan\mu_{n+1} = \frac{z + (\varepsilon^2 c)\sin^3\theta_n}{w - (e^2 a)\cos^3\theta_n} \qquad (2.51)$$

Altitude is determined as:

$$h = w\cos\mu + z\sin\mu - N(1 - e^2\sin^2\mu) \qquad (2.52)$$

Table 2.5 WGS84 Geometry.

WGS84 Parameter	Value
Equatorial radius (a)	6 378 137 m
First eccentricity (e)	0.081819218
Second eccentricity (ε)	0.082094465

2.4.2 Quaternions

Aircraft orientation is expressed using quaternions, which were explained in Volume 1. These parameters represent the rotation of an axis system about a line passing through its origin, as shown Figure 2.10. The rotation is defined by three direction angles (a, β, γ) and one rotation angle (δ).

The four quaternions are defined as a vector $\boldsymbol{e} = (e_0, e_1, e_2, e_3)^T$, where:

$$e_0 = \cos\frac{\delta}{2} \qquad e_1 = \cos a \sin\frac{\delta}{2} \qquad e_2 = \cos\beta \sin\frac{\delta}{2} \qquad e_3 = \cos\gamma \sin\frac{\delta}{2} \quad (2.53)$$

After a lot of manipulation, the aircraft orientation can be expressed by a direction cosine matrix:

$$R^{12} = \begin{pmatrix} e_0^2 + e_1^2 - e_2^2 - e_3^2 & 2(e_1 e_2 + e_3 e_0) & 2(e_3 e_1 - e_2 e_0) \\ 2(e_1 e_2 - e_3 e_0) & e_0^2 - e_1^2 + e_2^2 - e_3^2 & 2(e_2 e_3 + e_1 e_0) \\ 2(e_3 e_1 + e_2 e_0) & 2(e_2 e_3 - e_1 e_0) & e_0^2 - e_1^2 - e_2^2 + e_3^2 \end{pmatrix} \quad (2.54)$$

Azimuth, pitch, and roll angles (ψ, θ, φ) can be determined from this using Equation 2.12. The quaternion vector can be derived from azimuth, pitch, and roll angles as follows:

$$e_0 = \cos\frac{\psi}{2}\cos\frac{\theta}{2}\cos\frac{\varphi}{2} + \sin\frac{\psi}{2}\sin\frac{\theta}{2}\sin\frac{\varphi}{2} \qquad e_1 = \cos\frac{\psi}{2}\cos\frac{\theta}{2}\sin\frac{\varphi}{2} - \sin\frac{\psi}{2}\sin\frac{\theta}{2}\cos\frac{\varphi}{2}$$

$$e_2 = \cos\frac{\psi}{2}\sin\frac{\theta}{2}\cos\frac{\varphi}{2} + \sin\frac{\psi}{2}\cos\frac{\theta}{2}\sin\frac{\varphi}{2} \qquad e_3 = \sin\frac{\psi}{2}\cos\frac{\theta}{2}\cos\frac{\varphi}{2} - \cos\frac{\psi}{2}\sin\frac{\theta}{2}\sin\frac{\varphi}{2}$$

$$(2.55)$$

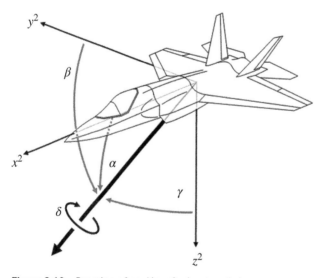

Figure 2.10 Rotation of an Aircraft about an Axis.

2.4.3 Kinematic Equations of Motion

Knowing the aircraft velocity in $\mathcal{F}^2$, the variation of aircraft position in $\mathcal{F}^0$ is obtained by a simple transformation:

$$\dot{r}^0 = R^{02}V^2 \tag{2.56}$$

Given angular velocity components (p, q, r), the rate of change of orientation is calculated as:

$$\dot{e} = \frac{1}{2}\begin{pmatrix} 0 & -p & -q & -r \\ p & 0 & -r & q \\ q & r & 0 & -p \\ r & -q & p & 0 \end{pmatrix}e \tag{2.57}$$

An alternative formulation exists (Nikravesh, 1988), which states that:

$$2R\dot{e} = \omega \tag{2.58}$$

where ω is the angular velocity vector and R is derived fom transformation matrix R^{21}, such that:

$$R^{21} = \mathcal{L}\mathcal{R}^T \tag{2.59}$$

where

$$\mathcal{L} = \begin{pmatrix} -e_1 & e_0 & -e_3 & e_2 \\ -e_2 & e_3 & e_0 & -e_1 \\ -e_3 & -e_2 & e_1 & e_0 \end{pmatrix} \qquad \mathcal{R} = \begin{pmatrix} -e_1 & e_0 & e_3 & -e_2 \\ -e_2 & -e_3 & e_0 & e_1 \\ -e_3 & e_2 & -e_1 & e_0 \end{pmatrix}$$

For clarity, the kinematics can be re-packaged as:

$$\dot{x}_{kin} = A_{kin}x_{kin} + B_{kin}u_{kin} \tag{2.60}$$

The state and input vectors are:

$$x_{kin} = \begin{pmatrix} r^0 \\ e^{02} \end{pmatrix} \qquad u_{kin} = \begin{pmatrix} V^2 \\ \omega^2 \end{pmatrix} \tag{2.61}$$

The state transition matrix A and the input matrix B are defined as:

$$A_{kin} = \begin{pmatrix} R^{02} & 0 \\ 0 & \mathcal{R}^T/2 \end{pmatrix} \qquad B_{kin} = \begin{pmatrix} I & 0 \\ 0 & I \end{pmatrix} \tag{2.62}$$

The dataflow associated with aircraft kinematics is summarised in Figure 2.11.

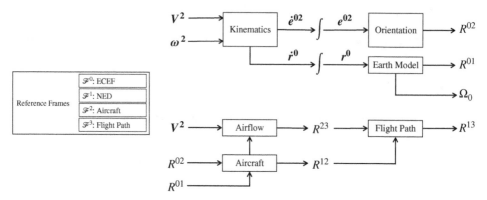

Figure 2.11 Dataflow for Kinematic Equations of Motion.

2.5 Initialisation

2.5.1 Balancing Forces

It is necessary to balance the forces and moments acting on an aircraft In order to establish its initial condition in flight. In this context, it is easiest to consider the aircraft to be a point mass flying through stationary air and then to establish the relevant force system with respect to flight path axes. In effect, forces and velocity are defined at the aircraft CG.

Figure 2.12 defines a generic flight condition with respect to the flight path, identifying Thrust T, Weight W, Lift force L and Drag force D. In addition, the velocity vector V is aligned with the instantaneous direction of travel and centripetal accelerations are shown in the vertical plane (associated with the turn rate $\dot{\gamma}_2$) and in the horizontal plane (associated with the turn rate $\dot{\gamma}_3$).

The equations of motion for airspeed and flight path axes were stated in Equation 1.4:

$$m\dot{V} = T - D - W \sin\gamma_2$$

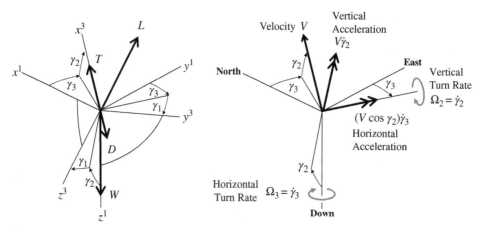

Figure 2.12 Generic Flight Condition.

$$mV\dot{\gamma}_2 = L\cos\gamma_1 - W\cos\gamma_2$$
$$mV\cos\gamma_2\dot{\gamma}_3 = L\sin\gamma_1$$

where $W = mg$. For equilibrium, airspeed (V) and turn rates ($\Omega_2 = \dot{\gamma}_2$ and $\Omega_3 = \dot{\gamma}_3$) are all predefined constants. So, the initial conditions are:

$$T = D + mg\sin\gamma_2$$

$$L = mV\frac{\Omega_2 + \dfrac{g}{V}\cos\gamma_2}{\cos\gamma_1} \tag{2.63}$$

$$\tan\gamma_1 = -\frac{\Omega_3\cos\gamma_2}{\Omega_2 + \dfrac{g}{V}\cos\gamma_2}$$

In addition, the load factor n is defined as lift divided by weight ($n = L/W$), such that:

$$n = \frac{\dfrac{g}{V}\Omega_2 + \cos\gamma_2}{\cos\gamma_1} \tag{2.64}$$

The rotation of the aircraft is driven by the vertical turn rate Ω_2 and the horizontal turn rate Ω_3, as shown in Figure 2.13, resolved with respect to frame $\mathcal{F}^1$. Also, the angular velocities (p, q, r) are resolved with respect to frame $\mathcal{F}^3$.

$$\omega^3 = R^{31}\omega^1 \tag{2.65}$$

where

$$\omega^1 = \begin{pmatrix} -\Omega_2\sin\gamma_3 \\ \Omega_2\cos\gamma_3 \\ \Omega_3 \end{pmatrix} \tag{2.66}$$

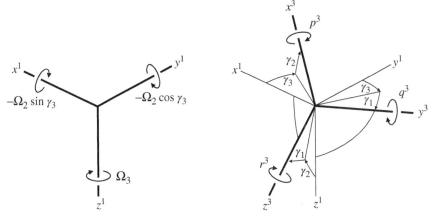

Figure 2.13 Angular Velocities for Flight Equilibrium.

Matrix R^{31} is the transpose of R^{13} in Equation 2.22:

$$R^{31} = \begin{pmatrix} \cos\gamma_2\cos\gamma_3 & \cos\gamma_2\sin\gamma_3 & -\sin\gamma_2 \\ \sin\gamma_1\sin\gamma_2\cos\gamma_3 - \cos\gamma_1\sin\gamma_3 & \sin\gamma_1\sin\gamma_2\sin\gamma_3 + \cos\gamma_1\cos\gamma_3 & \sin\gamma_1\cos\gamma_2 \\ \cos\gamma_1\sin\gamma_2\cos\gamma_3 + \sin\gamma_1\sin\gamma_3 & \cos\gamma_1\sin\gamma_2\sin\gamma_3 - \sin\gamma_1\cos\gamma_3 & \cos\gamma_1\cos\gamma_2 \end{pmatrix}$$

With a little effort, the steady-state angular velocity components are derived as:

$$\boldsymbol{\omega}^3 = \begin{pmatrix} -\Omega_3\sin\gamma_2 \\ \Omega_3\cos\gamma_2\sin\gamma_1 + \Omega_2\cos\gamma_1 \\ \Omega_3\cos\gamma_2\cos\gamma_1 - \Omega_2\sin\gamma_1 \end{pmatrix} \tag{2.67}$$

In addition, the force components are obtained as:

$$\boldsymbol{F}^3 = \begin{pmatrix} T - F_D - W\sin\gamma_2 \\ W\cos\gamma_2\sin\gamma_1 \\ W\cos\gamma_2\cos\gamma_1 - F_L \end{pmatrix} \tag{2.68}$$

2.5.2 Typical Flight Conditions

In *straight-line* flight, the turn rates Ω_2 and Ω_3 are identically zero. Thus, for a constant climb/dive angle, γ_2:

$$\gamma_1 = 0 \qquad T = D + mg\sin\gamma_2 \qquad n = \cos\gamma_2 \qquad \boldsymbol{\omega}^3 = \begin{pmatrix} 0 \\ 0 \\ 0 \end{pmatrix} \tag{2.69}$$

In *straight-and-level* flight ($\gamma_2 = 0$), equilibrium is achieved when thrust equals drag and lift equal weight:

$$\gamma_1 = 0 \qquad T = D \qquad n = 1 \qquad \boldsymbol{\omega}^3 = \begin{pmatrix} 0 \\ 0 \\ 0 \end{pmatrix} \tag{2.70}$$

For a *sustained turn* at a constant climb/dive angle with turn rates $\Omega_2 = 0$ and $\Omega_3 \neq 0$, equilibrium is achieved as follows:

$$\tan\gamma_1 = -\frac{V}{g}\Omega_3 \qquad T = D + mg\sin\gamma_2 \qquad n = \frac{\cos\gamma_2}{\cos\gamma_1} \qquad \boldsymbol{\omega}^3 = \begin{pmatrix} -\Omega_3\sin\gamma_2 \\ \Omega_3\cos\gamma_2\sin\gamma_1 \\ \Omega_3\cos\gamma_2\cos\gamma_1 \end{pmatrix} \tag{2.71}$$

For a *sustained horizontal turn* (where $\gamma_2 = 0$), this is simplified, as follows:

$$\tan \gamma_1 = -\frac{V}{g}\Omega_3 \qquad T = D \qquad n = \frac{1}{\cos \gamma_1} \qquad \omega^3 = \begin{pmatrix} 0 \\ \Omega_3 \sin \gamma_1 \\ \Omega_3 \cos \gamma_1 \end{pmatrix}$$

(2.72)

A *vertical pull-up* is defined with turn rates $\Omega_2 \neq 0$ and $\Omega_3 = 0$. Instantaneously, the trajectory is a circular path and a snapshot is taken at the bottom of the turn such that $\gamma_2 = 0$. This is NOT a steady-state condition but flight parameters can still be determined:

$$\gamma_1 = 0 \qquad T = D \qquad n = 1 + \frac{V}{g}\Omega_2 \qquad \omega^3 = \begin{pmatrix} 0 \\ \Omega_2 \\ 0 \end{pmatrix}$$

(2.73)

2.5.3 Finding Aircraft Flight Parameters for Equilibrium

An equilibrium condition is defined by airspeed (V), track angle (γ_3), climb/dive angle (γ_2), vertical turn rate (Ω_2), and horizontal turn rate (Ω_3). These parameters allow the calculation of bank angle (γ_1), load factor (n), angular velocity (ω^3), and nett propulsive force ($T - D$), as shown in the Section 2.5.2. All vector quantities are defined at the CG and, to avoid confusion, this will be denoted by a subscript G in what follows.

The linear velocity along the flight path is defined as:

$$V_G^3 = \begin{pmatrix} V \\ 0 \\ 0 \end{pmatrix}$$

(2.74)

The angular velocity vector ω_G^3 is defined by Equation 2.68 and the force vector F_G^3 is defined by Equation 2.69. By definition, the moment vector about the CG is zero ($M_G^4 = 0$) because all forces act through the CG.

These vectors are transformed from flight path to aircraft by using matrix R^{24} from Equation 2.25:

$$\begin{pmatrix} V \\ \omega \end{pmatrix}_G^2 = \begin{pmatrix} R^{24} & 0 \\ 0 & R^{24} \end{pmatrix} \begin{pmatrix} V \\ \omega \end{pmatrix}_G^3 \qquad \begin{pmatrix} F \\ M \end{pmatrix}_G^2 = \begin{pmatrix} R^{24} & 0 \\ 0 & R^{24} \end{pmatrix} \begin{pmatrix} F \\ M \end{pmatrix}_G^3$$

(2.75)

where

$$R^{23} = \begin{pmatrix} \cos \alpha \cos \beta & -\cos \alpha \sin \beta & -\sin \alpha \\ \sin \beta & \cos \beta & 0 \\ \sin \alpha \cos \beta & -\sin \alpha \sin \beta & \cos \alpha \end{pmatrix}$$

Recalling Equation 2.37:

$$\begin{pmatrix} F \\ M \end{pmatrix} = \begin{pmatrix} I & 0 \\ -G & I \end{pmatrix} \begin{pmatrix} F \\ M \end{pmatrix}_G \qquad \begin{pmatrix} V \\ \omega \end{pmatrix} = \begin{pmatrix} I & -G \\ 0 & I \end{pmatrix} \begin{pmatrix} V \\ \omega \end{pmatrix}_G$$

Thus, these vectors can be mapped from the CG to the aircraft datum, as follows:

$$\begin{pmatrix} V \\ \omega \end{pmatrix}^2 = \begin{pmatrix} I & -G \\ 0 & I \end{pmatrix}\begin{pmatrix} V \\ \omega \end{pmatrix}_G^2 \qquad \begin{pmatrix} F \\ M \end{pmatrix}^2 = \begin{pmatrix} I & 0 \\ -G & I \end{pmatrix}\begin{pmatrix} F \\ M \end{pmatrix}_G^2 \qquad (2.76)$$

This sets up the initialisation if and only matrix R^{23} is known, which means that AOA (α) and AOS (β) need to be found [cf. Figure 2.6]. This requires insight from other chapters. For an aircraft at a given altitude, forces and moments can be written as functions:

$$F^2 = f\big(V, \alpha, \beta, \dot\alpha, \dot\beta, p, q, r, \theta, \varphi, T, \delta_E, \delta_A, \delta_R\big)$$

$$M^2 = g\big(V, \alpha, \beta, \dot\alpha, \dot\beta, p, q, r, \theta, \varphi, T, \delta_E, \delta_A, \delta_R\big)$$

where

$$F^2 = \begin{pmatrix} X \\ Y \\ Z \end{pmatrix} \qquad M^2 = \begin{pmatrix} L \\ M \\ N \end{pmatrix}$$

Airspeed (V) is constant. Angular velocities (p, q, r) are predefined. Pitch and roll angles (θ, φ) are constrained by the combination of flight path angles (via matrix R^{13}) and AOA and AOS (via matrix R^{32}). The flight path angles are predefined and so, for initialisation, changes in α and β (as independent variables) will drive changes in θ and φ (as dependent variables). Also, rates of change $\dot\alpha$ and $\dot\beta$ are both zero because a steady state is being established.

Installed thrust is included as T. Flight controls are included as (δ_E, δ_A, δ_R), denoting elevator, aileron, and rudders deflections for pitch, roll, and yaw control, respectively.

With this in mind, forces and moments can be rewritten, as follows:

$$F^2 = f_0(V, p, q, r) + f(\alpha, \beta, T, \delta_E, \delta_A, \delta_R) \qquad (2.77)$$

$$M^2 = g_0(V, p, q, r) + g(\alpha, \beta, T, \delta_E, \delta_A, \delta_R) \qquad (2.78)$$

where the expressions are split between constants and independent variables. The dependent variables and the 'zero' contributors have been removed explicitly.

Invariably, the initialisation process is iterative, driven by an input vector $\boldsymbol{u}$ and an output vector $\boldsymbol{y}$:

$$\boldsymbol{u} = \begin{pmatrix} T \\ \beta \\ \alpha \\ \delta_A \\ \delta_E \\ \delta_R \end{pmatrix} \qquad \boldsymbol{y} = \begin{pmatrix} X \\ Y \\ Z \\ L \\ M \\ N \end{pmatrix} \qquad (2.79)$$

and a force/moment calculation:

$$\boldsymbol{y} = F(\boldsymbol{u}) \qquad (2.80)$$

The process is seeded with $\boldsymbol{u} = \boldsymbol{u}_0$, in which all inputs are zero, and the resulting output $\boldsymbol{y}_0 = F(\boldsymbol{u}_0)$ establishes default values for the force/moment components:

$$
\boldsymbol{u}_0 = \begin{pmatrix} 0 \\ 0 \\ 0 \\ 0 \\ 0 \\ 0 \end{pmatrix} \longrightarrow \boldsymbol{y}_0 = \begin{pmatrix} X_0 \\ Y_0 \\ Z_0 \\ L_0 \\ M_0 \\ N_0 \end{pmatrix}
$$

Generally, the quickest method to employ is a gradient search. For each iteration, the change in output is divided by the change in input and then this ratio is used to decide what the input should be for the next iteration. The process starts with output $\boldsymbol{y}_0$ and continues with $\boldsymbol{y}_1, \boldsymbol{y}_2, \boldsymbol{y}_3, \ldots$ and so on until each output variable is less than a predefined threshold. In other words, the objective is for $\boldsymbol{y}_n \longrightarrow \boldsymbol{0}$ as n increases (with the practical expection that n should not become too large!).

For the first iteration ($n = 1$), a fixed increment $\Delta \boldsymbol{u}_1$ is applied to the starting input $\boldsymbol{u}_0$:

$$
\boldsymbol{u}_0 \longrightarrow \boldsymbol{y}_0 \longrightarrow \boldsymbol{u}_1 = \boldsymbol{u}_0 + \Delta \boldsymbol{u}_1 \longrightarrow \boldsymbol{y}_1 = F(\boldsymbol{u}_1)
$$

Gradients are calculated for the six input/output pairings ($k = 1, 2, \ldots, 6$):

$$
g_{k,1} = \frac{y_{k,1} - y_{k,0}}{x_{k,1} - x_{k,0}} \tag{2.81}
$$

For subsequent iterations ($n = 2, 3, 4, \ldots$), the incremental changes to inputs are calculated as:

$$
\Delta u_{k,n} = -\frac{y_{k,n-1}}{g_{k,n-1}} \tag{2.82}
$$

An increment $\Delta \boldsymbol{u}_n$ is applied to the previous input $\boldsymbol{u}_{n-1}$:

$$
\boldsymbol{u}_{n-1} \longrightarrow \boldsymbol{y}_{n-1} \longrightarrow \boldsymbol{u}_n = \boldsymbol{u}_{n-1} + \Delta \boldsymbol{u}_n \longrightarrow \boldsymbol{y}_n = F(\boldsymbol{u}_n)
$$

Gradients are now calculated for the six input/output pairings ($k = 1, 2, \ldots, 6$):

$$
g_{k,n} = \frac{y_{k,n} - y_{k,n-1}}{x_{k,n} - x_{k,n-1}} \tag{2.83}
$$

The iterations continue until convergence is achieved. If there is no convergence, then something is wrong in the force/moment calculation. It is noted that convergence after N iterations means that:

$$
\begin{pmatrix} X_N \\ Y_N \\ Z_N \\ L_N \\ M_N \\ N_N \end{pmatrix} \longrightarrow \begin{pmatrix} 0 \\ 0 \\ 0 \\ 0 \\ 0 \\ 0 \end{pmatrix} \tag{2.84}
$$

2.6 Linearisation

Once the aircraft has been initialised, a linearised model can be created that describes motion in the vicinity of the initial condition. This encompasses the equations of motion and the force/moment system. The end result is a set of matrix equations that can be subjected to linear analysis.

Linearisation gives an idealised snapshot of the aircraft dynamics, such that the initial condition can be sustained forever (which is called a *trim condition*). All off-trim conditions are transient excursions that are 'small'. In particular, this implies that mass properties are held constant.

For heuristic purposes, the force/moment and velocity vector are usually referred to the CG, such that $G = 0$. This will be done implicitly, without use of subscripts, in order to avoid the algebra becoming cluttered.

Invariably, linearisation produces a flight model that is referred to the local NED reference frame $\mathcal{F}^1$, not the ECEF frame $\mathcal{F}^0$. In addition, because linear analysis is only performed for flight segments of short range and short duration, the rotation and curvature of the Earth are ignored. This means that $\Omega \longrightarrow \Omega_\omega$.

2.6.1 Linearisation of Dynamic Equations of Motion

Equation 2.39 defines the generalised form of the dynamic equations of motion:

$$\begin{pmatrix} m & mG \\ 0 & J \end{pmatrix} \begin{pmatrix} \dot{V} \\ \dot{\omega} \end{pmatrix} = \begin{pmatrix} I & 0 \\ G & I \end{pmatrix} \begin{pmatrix} F \\ M \end{pmatrix} - \begin{pmatrix} \dot{m} + \Omega m & (\dot{m} + \Omega m)G + m\dot{G} \\ 0 & \dot{J} + \Omega J \end{pmatrix} \begin{pmatrix} V \\ \omega \end{pmatrix}$$

These are simplified, as:

$$\begin{pmatrix} m & 0 \\ 0 & J \end{pmatrix} \begin{pmatrix} \dot{V} \\ \dot{\omega} \end{pmatrix} = \begin{pmatrix} I & 0 \\ 0 & I \end{pmatrix} \begin{pmatrix} F \\ M \end{pmatrix} - \begin{pmatrix} \Omega_\omega m & 0 \\ 0 & \Omega_\omega J \end{pmatrix} \begin{pmatrix} V \\ \omega \end{pmatrix} \tag{2.85}$$

From here, the vector set (V, ω, F, M) is replaced with $(V + \Delta V, \omega + \Delta \omega, F + \Delta F, M + \Delta M)$, where the additional terms are small perturbations:

$$\Delta F = \begin{pmatrix} \Delta X^1 \\ \Delta Y^1 \\ \Delta Z^1 \end{pmatrix} \qquad \Delta M = \begin{pmatrix} \Delta L^1 \\ \Delta M^1 \\ \Delta N^1 \end{pmatrix} \qquad \Delta V = \begin{pmatrix} \Delta u^1 \\ \Delta v^1 \\ \Delta w^1 \end{pmatrix} \qquad \Delta \omega = \begin{pmatrix} \Delta p^1 \\ \Delta q^1 \\ \Delta r^1 \end{pmatrix}$$

$$\tag{2.86}$$

In addition, it is necessary to replace Ω with $\Omega + \Delta \Omega$, where:

$$\Delta \Omega = \begin{pmatrix} 0 & -\Delta r & \Delta q \\ \Delta r & 0 & -\Delta p \\ -\Delta q & \Delta p & 0 \end{pmatrix} \tag{2.87}$$

Making these replacements:

$$\begin{pmatrix} m & 0 \\ 0 & J \end{pmatrix}\begin{pmatrix} \dot{V}+\Delta\dot{V} \\ \dot{\omega}+\Delta\dot{\omega} \end{pmatrix} = \begin{pmatrix} I & 0 \\ G & I \end{pmatrix}\begin{pmatrix} F+\Delta F \\ M+\Delta M \end{pmatrix} - \begin{pmatrix} (\Omega_\omega+\Delta\Omega_\omega)m & 0 \\ 0 & (\Omega_\omega+\Delta\Omega_\omega)J \end{pmatrix}\begin{pmatrix} V+\Delta V \\ \omega+\Delta\omega \end{pmatrix}$$

Now, subtract Equation 2.85 to derive the linearised equations of motion:

$$\begin{pmatrix} m & 0 \\ 0 & J \end{pmatrix}\begin{pmatrix} \Delta\dot{V} \\ \Delta\dot{\omega} \end{pmatrix} = \begin{pmatrix} I & 0 \\ 0 & I \end{pmatrix}\begin{pmatrix} \Delta F \\ \Delta M \end{pmatrix} - \begin{pmatrix} \Delta\Omega_\omega m & 0 \\ 0 & \Delta\Omega_\omega J \end{pmatrix}\begin{pmatrix} V \\ \omega \end{pmatrix} - \begin{pmatrix} \Omega_\omega m & 0 \\ 0 & \Omega_\omega J \end{pmatrix}\begin{pmatrix} \Delta V \\ \Delta\omega \end{pmatrix} \tag{2.88}$$

Importantly, the products of small perturbations are ignored because they give rise to even smaller perturbations.

A small amount of re-arrangement is needed in order to remove the explicitly dependency on V and ω. It is noted that:

$$\Delta\Omega_\omega(mV) = \Delta\omega \times (mV) \qquad\qquad \Delta\Omega_\omega(J\omega) = \Delta\omega \times (J\omega)$$

These expressions are equivalent to:

$$\Delta\Omega_\omega(mV) = -(mV) \times \Delta\omega \qquad\qquad \Delta\Omega_\omega(J\omega) = -(J\omega) \times \Delta\omega$$
$$\Delta\Omega_\omega(mV) = -m\Omega_V\Delta\omega \qquad\qquad \Delta\Omega_\omega(J\omega) = -J\Omega_\omega\Delta\omega$$

where the cross-product matrices are:

$$\Omega_V = \begin{pmatrix} 0 & -v & w \\ w & 0 & -u \\ -v & u & 0 \end{pmatrix} \qquad\qquad \Omega_\omega = \begin{pmatrix} 0 & -r & q \\ r & 0 & -p \\ -q & p & 0 \end{pmatrix} \tag{2.89}$$

This modifies Equation 2.88, as follows:

$$\begin{pmatrix} m & 0 \\ 0 & J \end{pmatrix}\begin{pmatrix} \Delta\dot{V} \\ \Delta\dot{\omega} \end{pmatrix} = \begin{pmatrix} I & 0 \\ 0 & I \end{pmatrix}\begin{pmatrix} \Delta F \\ \Delta M \end{pmatrix} - \begin{pmatrix} \Omega_\omega m & -m\Omega_V \\ 0 & \Omega_\omega J - J\Omega_\omega \end{pmatrix}\begin{pmatrix} \Delta V \\ \Delta\omega \end{pmatrix} \tag{2.90}$$

2.6.2 Linearisation of Kinematic Equations of Motion

Now, variations of aircraft velocity in $\mathcal{F}^2$ cause variations in aircraft position in $\mathcal{F}^1$:

$$\Delta\dot{r} = R^{01}\Delta V \tag{2.91}$$

Small changes in orientation are measured using Euler angles (azimuth, pitch, and roll). It is convenient to package these into an aircraft 'attitude' vector:

$$a = \begin{pmatrix} \varphi \\ \theta \\ \psi \end{pmatrix} \tag{2.92}$$

For heuristic purposes, textbooks (including this one) assume a straight-and-level trim condition at high speed, for which the pitch and roll angles can be neglected. In this case, the rate of change of orientation is expressed in the simplest possible form:

$$\Delta \dot{a} = \begin{pmatrix} 1 & 0 & 0 \\ 0 & 1 & 0 \\ 0 & 0 & 1 \end{pmatrix} \Delta \omega \qquad (2.93)$$

This avoids a lengthy derivation of the general case. If this were needed, then Volume 1 presents kinematic equations expressed using Euler angles. These can be linearised using essentially the same method that was applied to the dynamic equations in Section 2.7.1.

The consolidated kinematic equations are:

$$\begin{pmatrix} \Delta \dot{r} \\ \Delta \dot{a} \end{pmatrix} = \begin{pmatrix} R^{01} & 0 \\ 0 & I \end{pmatrix} \begin{pmatrix} \Delta V \\ \Delta \omega \end{pmatrix} \qquad (2.94)$$

2.6.3 Linearisation of Aerodynamic Forces and Moments

In its most general formulation, the aerodynamic force/moment system is dependent on linear velocity V and angular velocity ω and their rates of change, as well as control deflections δ. Altitude affects dynamic pressure and is obtained from the position vector r. This can be written as:

$$F_A = f(V, \dot{V}, \omega, \dot{\omega}, r, \delta)$$
$$M_A = g(V, \dot{V}, \omega, \dot{\omega}, r, \delta)$$

where

$$F_A = \begin{pmatrix} X_A \\ Y_A \\ Z_A \end{pmatrix} \qquad M_A = \begin{pmatrix} L_A \\ M_A \\ N_A \end{pmatrix} \qquad V = \begin{pmatrix} u \\ v \\ w \end{pmatrix} \qquad \omega = \begin{pmatrix} p \\ q \\ r \end{pmatrix} \qquad r = \begin{pmatrix} n \\ e \\ d \end{pmatrix}$$

Small perturbations can be expressed as follows:

$$\begin{pmatrix} \Delta F_A \\ \Delta M_A \end{pmatrix} = \begin{pmatrix} F_V & F_\omega \\ M_V & M_\omega \end{pmatrix} \begin{pmatrix} \Delta V \\ \Delta \omega \end{pmatrix} + \begin{pmatrix} F_{\dot{V}} & F_{\dot{\omega}} \\ M_{\dot{V}} & M_{\dot{\omega}} \end{pmatrix} \begin{pmatrix} \Delta \dot{V} \\ \Delta \dot{\omega} \end{pmatrix} + \begin{pmatrix} F_\delta \\ M_\delta \end{pmatrix} \Delta \delta + \begin{pmatrix} F_r \\ M_r \end{pmatrix} \Delta r \qquad (2.95)$$

The matrix elements are Jacobian matrices of the form:

$$F_V = \frac{\partial F}{\partial V} = \begin{pmatrix} \dfrac{\partial X_A}{\partial u} & \dfrac{\partial X_A}{\partial v} & \dfrac{\partial X_A}{\partial w} \\ \dfrac{\partial Y_A}{\partial u} & \dfrac{\partial Y_A}{\partial v} & \dfrac{\partial Y_A}{\partial w} \\ \dfrac{\partial Z_A}{\partial u} & \dfrac{\partial Z_A}{\partial v} & \dfrac{\partial Z_A}{\partial w} \end{pmatrix} \qquad (2.96)$$

All other elements are defined similarly.

2.6.4 Linearisation of Propulsive Forces and Moments

A simple propulsion system concept would comprise a single thrust generator that is contained in the plane of symmetry and is aligned with the horizontal aircraft datum. The thrust line passes through a propulsion reference point $P = (x_P, 0, z_P)$. Thus, the representation of propulsive forces and moments can be idealised as follows:

$$F_P = \begin{pmatrix} 1 \\ 0 \\ 0 \end{pmatrix} T \qquad M_P = P \times F_P = \Omega_P F_P \tag{2.97}$$

where Ω_P is the cross-product matrix:

$$\Omega_P = \begin{pmatrix} 0 & -z_P & 0 \\ z_P & 0 & -x_P \\ 0 & x_P & 0 \end{pmatrix} \tag{2.98}$$

The linearised form of these equations can be written as:

$$\begin{pmatrix} \Delta F_P \\ \Delta M_P \end{pmatrix} = \begin{pmatrix} F_P \\ M_P \end{pmatrix} \Delta T \tag{2.99}$$

where

$$F_P = \begin{pmatrix} 1 \\ 0 \\ 0 \end{pmatrix} \qquad M_P = \Omega_P \begin{pmatrix} 1 \\ 0 \\ 0 \end{pmatrix}$$

2.6.5 Linearisation of Gravitational Forces and Moments

Gravitational force and moment components are defined by:

$$F_G = R^{21} g^1 = mg \begin{pmatrix} -\sin\theta \\ \sin\varphi\cos\theta \\ \cos\varphi\cos\theta \end{pmatrix} \qquad M_G = \Omega_G F_G \tag{2.100}$$

where the trigonometry comes from Equation 2.19 and the cross-product matrix $\Omega_G \equiv G$ is defined in Equation 2.39.

Small perturbations are added as follows:

$$F_G + \Delta F_G = mg \begin{pmatrix} -\sin(\theta + \Delta\theta) \\ \sin(\varphi + \Delta\varphi)\cos(\theta + \Delta\theta) \\ \cos(\varphi + \Delta\varphi)\cos(\theta + \Delta\theta) \end{pmatrix} \qquad M_G + \Delta M_G = \Omega_r(F_G + \Delta F_G)$$

Noting that $\Delta\theta$ and $\Delta\varphi$ are small angles, trigonometric expansions are performed as:

$$\sin(\theta + \Delta\theta) \approx \sin\theta + \Delta\theta\cos\theta \qquad \cos(\theta + \Delta\theta) \approx \cos\theta - \Delta\theta\sin\theta$$

and

$$\sin(\varphi + \Delta\varphi) \approx \sin\varphi + \Delta\varphi\cos\varphi \qquad \cos(\varphi + \Delta\varphi) \approx \cos\varphi - \Delta\varphi\sin\varphi$$

The perturbations in F_G and M_G are then given by:

$$\Delta F_G = mg\begin{pmatrix} -\Delta\theta\cos\theta \\ \Delta\varphi\cos\varphi\cos\theta - \Delta\theta\sin\varphi\sin\theta \\ -\Delta\varphi\sin\varphi\cos\theta - \Delta\theta\cos\varphi\sin\theta \end{pmatrix} \qquad \Delta M_G = \Omega_r\Delta F_G$$

This can be written in a compact form:

$$\begin{pmatrix} \Delta F_G \\ \Delta M_G \end{pmatrix} = \begin{pmatrix} F_G \\ M_G \end{pmatrix}\Delta a \tag{2.101}$$

where

$$F_G = mg\begin{pmatrix} 0 & -\cos\theta & 0 \\ \cos\varphi\cos\theta & -\sin\varphi\sin\theta & 0 \\ -\sin\varphi\cos\theta & -\cos\varphi\sin\theta & 0 \end{pmatrix} \qquad M_G = GF_G$$

The attitude vector is defined in Equation 2.92. Its perturbation is:

$$\Delta a = \begin{pmatrix} \Delta\varphi \\ \Delta\theta \\ \Delta\psi \end{pmatrix}$$

As stated already, products of small perturbations are ignored because they are very small.

2.6.6 The Complete Linearised System of Equations

Dynamic equations of motion are given in Equation 2.90:

$$\begin{pmatrix} m & 0 \\ 0 & J \end{pmatrix}\begin{pmatrix} \Delta\dot{V} \\ \Delta\dot{\omega} \end{pmatrix} = \begin{pmatrix} I & 0 \\ 0 & I \end{pmatrix}\begin{pmatrix} \Delta F \\ \Delta M \end{pmatrix} - \begin{pmatrix} \Omega_\omega m & -m\Omega_V \\ 0 & \Omega_\omega J - J\Omega_\omega \end{pmatrix}\begin{pmatrix} \Delta V \\ \Delta\omega \end{pmatrix}$$

Aerodynamic forces and moments are given in Equation 2.95:

$$\begin{pmatrix} \Delta F_A \\ \Delta M_A \end{pmatrix} = \begin{pmatrix} F_V & F_\omega \\ M_V & M_\omega \end{pmatrix}\begin{pmatrix} \Delta V \\ \Delta\omega \end{pmatrix} + \begin{pmatrix} F_{\dot{V}} & F_{\dot{\omega}} \\ M_{\dot{V}} & M_{\dot{\omega}} \end{pmatrix}\begin{pmatrix} \Delta\dot{V} \\ \Delta\dot{\omega} \end{pmatrix} + \begin{pmatrix} F_\delta \\ M_\delta \end{pmatrix}\Delta\delta + \begin{pmatrix} F_r \\ M_r \end{pmatrix}\Delta r$$

Propulsive forces and moments are given in Equation 2.99:

$$\begin{pmatrix} \Delta F_P \\ \Delta M_P \end{pmatrix} = \begin{pmatrix} F_P \\ M_P \end{pmatrix}\Delta T$$

Gravitational forces and moments are given in Equation 2.101:

$$\begin{pmatrix} \Delta F_G \\ \Delta M_G \end{pmatrix} = \begin{pmatrix} F_G \\ M_G \end{pmatrix}\Delta a$$

Forces and moments are combined as:

$$\begin{pmatrix} \Delta F \\ \Delta M \end{pmatrix} = \begin{pmatrix} \Delta F_A \\ \Delta M_A \end{pmatrix} + \begin{pmatrix} \Delta F_P \\ \Delta M_P \end{pmatrix} + \begin{pmatrix} \Delta F_G \\ \Delta M_G \end{pmatrix} \tag{2.102}$$

Kinematic equations of motion are given in Equation 2.94:

$$\begin{pmatrix} \Delta \dot{r} \\ \Delta \dot{a} \end{pmatrix} = \begin{pmatrix} R^{01} & 0 \\ 0 & I \end{pmatrix} \begin{pmatrix} \Delta V \\ \Delta \omega \end{pmatrix}$$

After some manipulation, the linearised equations are consolidated as follows:

$$E \Delta \dot{x} = A \Delta x + B \Delta u \tag{2.103}$$

where

$$\Delta x = \begin{pmatrix} \Delta V \\ \Delta \omega \\ \Delta r \\ \Delta a \end{pmatrix} \qquad \Delta u = \begin{pmatrix} \Delta \delta \\ \Delta T \end{pmatrix} \tag{2.104}$$

The 'mass matrix' E, the state transition matrix A, and the control matrix B are defined as:

$$E = \begin{pmatrix} m - F_{\dot{V}} & -F_{\dot{\omega}} & 0 & 0 \\ -M_{\dot{V}} & J - M_{\dot{\omega}} & 0 & 0 \\ 0 & 0 & I & 0 \\ 0 & 0 & 0 & I \end{pmatrix} \tag{2.105a}$$

$$A = \begin{pmatrix} F_V - \Omega_\omega m & F_\omega - m\Omega_V & F_r & F_G \\ M_V & M_\omega - J\Omega_\omega - \Omega_\omega J & M_r & M_G \\ R^{01} & 0 & 0 & 0 \\ 0 & i & 0 & 0 \end{pmatrix} \qquad B = \begin{pmatrix} F_\delta & F_P \\ M_\delta & M_P \\ 0 & 0 \\ 0 & 0 \end{pmatrix} \tag{2.105b}$$

3

Fixed-Wing Aerodynamics

3.1 Introduction

3.1.1 Fixed Wings and Aerodynamics

Fixed-wing aerodynamics is probably the most widely discussed and explained topic in the aeronautical literature (including innumerable online sources). It is the technical basis for the vast majority of air vehicles and it has the virtue of being easy to visualise and, therefore, easy to understand. The handbook approach has been developed over many, many decades and has been distilled down to small number of methods for calculating lift, drag, and pitching moment. Accordingly, it is ideal for introductory courses in any undergraduate programme. However, typical course content might not cover practical methods of calculation (especially for drag) and, almost certainly, it will not deal with aerodynamic load distributions. So, this chapter will cover typical course content plus these additional topics.

3.1.2 What Chapter 3 Includes

This chapter includes:

- Aerodynamic Principles
  (covering aerofoils, force/moment definitions, aerodynamic centre, and wing geometry)
- Aerodynamic Model of an Isolated Wing
  (defining methods for calculating lift, drag, and pitching moment)
- Trailing-Edge Controls
  (defining methods for calculating incremental aerodynamics and hinge moments for plain flaps)
- Factors affecting Lift Generation
  (focusing on sideslip, aircraft rotation, structural flexibility, ground effect, and indicial effects)
- Lift Distribution
  (specifying Diederich's Method)
- Drag Distribution
  (summarising the mathematical principles of Lifting Line Theory)

3.1.3 What Chapter 3 Excludes

This chapter excludes:

- Aerodynamic interaction or interference between a wing and anything on the aircraft.
- Leading-edge controls.
- Trailing-edge controls other than plain flaps.

3.1.4 Overall Aim

Chapter 3 should provide a reasonably comprehensive set of methods for estimating the lift, drag, and pitching moment of a wing up to a Mach number of 0.85, plus Diederich's method for estimating the spanwise lift distribution. In addition, there is discussion on additional factors that influence lift generation and on an approach to estimating the drag distribution.

3.2 Aerodynamic Principles

3.2.1 Aerofoils

The aim of 'useful aerodynamics' is to generate sufficient lift in order to get a vehicle into the air and to keep it there. This means doing work on the air by propelling a wing at speed, either using thrust for powered flight or using gravity for gliding flight. This is achieved most efficiently using aerofoils, which are familiar curved shapes, rounded at the leading edge and tapering to a point at the trailing edge. The external flow pattern around a wing looks something like that shown in Figure 3.1.

The angle between the wing and the freestream is called the Angle of attack (AOA). Airflow is deflected upwards ahead of the leading edge (known as 'upwash') and deflected downwards behind the trailing edge (known as 'downwash'). Upper and lower streamlines separate at the stagnation point, at which the flow velocity is zero. Flow curvature and flow velocity are greater over the upper surface than the lower surface and the resulting pressure difference between the two surfaces generates lift.

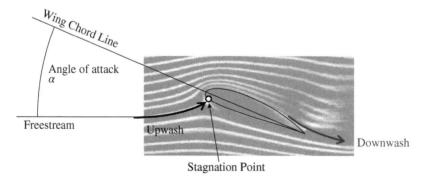

Figure 3.1 Streamlines around a Wing Section.

A parallel interpretation says that lift is equal to the change in vertical momentum, as the horizontal flow ahead of the wing is deflected downwards from the trailing edge. Alternatively, when streamlines are attached to the wing surfaces, the attachment is achieved by suction that acts on the air as a centripetal force. It can then be argued that a force on the air in one direction implies an equal and opposite force acting on the wing surface. Fundamentally, what is always true is that lift requires an asymmetric flow around the wing and the by-product is downwash.

A typical pressure distribution is shown in Figure 3.2. At any position, the pressure acts normal to the wing surface. This is depicted as force vectors, with components resolved in vertical and horizontal directions. Vertical force components add up to give the total lift: horizontal components add up to give the total drag.

The aerofoil cross-section can be symmetric or asymmetric, which is described as *uncambered* or *cambered*, respectively. This is shown in Figure 3.3. An uncambered section has no lift when the chord line is parallel with the airstream. The mean line between upper and lower surfaces is identically the chord line. A cambered wing has a mean line that is curved. This asymmetry produces lift even when the chord line is parallel with the airstream and the aerofoil must be rotated leading-edge down in order to achieve zero lift. In this condition, it is convenient to think of the chord line as being aligned with a zero-lift line. Also, the flow around a cambered wing is not symmetrical about the chord line, such that the lift vectors acting on the upper and lower surfaces are not aligned. Therefore, a pitching moment is generated. Typical pressure distributions are sketched in Figure 3.4.

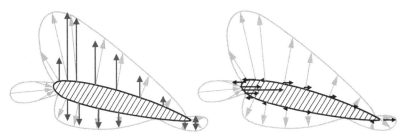

Figure 3.2 Pressure Distribution over a Wing Section.

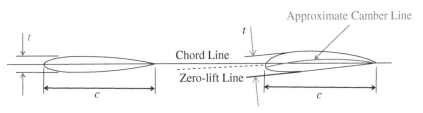

Figure 3.3 Uncambered and Cambered Wing Sections.

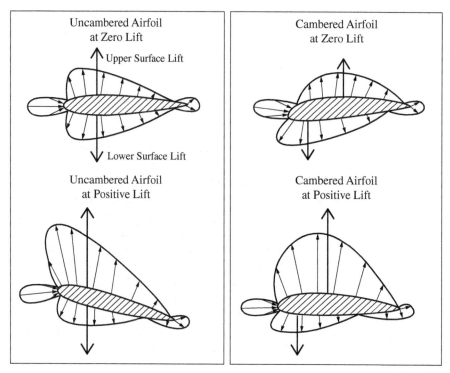

Figure 3.4 Forces on Uncambered and Cambered Wings.

3.2.2 Dimensional Analysis

Generically, an aerodynamic force F (i.e. lift or drag) depends on airspeed V, AOA α, and chord length c (measured from leading-edge to trailing-edge), as well as on air density ρ, dynamic viscosity μ and speed of sound a. This functional inter-dependence can be written as:

$$f(F, V, \alpha, c, \rho, \mu, a) = 0 \tag{3.1}$$

These variables have dimensions that are expressed as follows:

$$[F] = \mathrm{MLT}^{-2} \quad [V] = \mathrm{M}^0\mathrm{L}^1\mathrm{T}^{-1} \quad [\alpha] = \mathrm{M}^0\mathrm{L}^0\mathrm{T}^0 \quad [c] = \mathrm{M}^0\mathrm{L}^1\mathrm{T}^0$$

$$[\rho] = \mathrm{M}^1\mathrm{L}^{-3}\mathrm{T}^0 \quad [\mu] = \mathrm{M}^1\mathrm{L}^{-1}\mathrm{T}^{-1} \quad [a] = \mathrm{M}^0\mathrm{L}^1\mathrm{T}^{-1}$$

where the primary dimensions are mass M, length L and time T. Applying the Buckingham Pi method, choose three independent variables (V, c, ρ) that are dimensional (i.e. not non-dimensional). Next, define three nondimensional groups:

$$\Pi_1 = V^{i_1}c^{j_1}\rho^{k_1}F \quad \Pi_2 = V^{i_2}c^{j_2}\rho^{k_2}\mu \quad \Pi_3 = V^{i_3}c^{j_3}\rho^{k_3}a$$

such that:

$$[\Pi_1] = \mathrm{M}^0\mathrm{L}^0\mathrm{T}^0 \quad [\Pi_2] = \mathrm{M}^0\mathrm{L}^0\mathrm{T}^0 \quad [\Pi_3] = \mathrm{M}^0\mathrm{L}^0\mathrm{T}^0$$

Analysis proceeds by incorporating the dimensionality of the individual variables and then equating powers of M, L, and T. Thus, the indices $i_n, j_n,$ and k_n are evaluated for each group, as follows:

$$\Pi_1 = \frac{F}{\rho V^2 c^2} \qquad \Pi_2 = \frac{\mu}{\rho V c} \qquad \Pi_3 = \frac{a}{V}$$

Without loss of generality, these groups can be redefined:

$$\Pi_1 \triangleq \frac{F}{\frac{1}{2}\rho V^2 S} \qquad \Pi_2 \triangleq \frac{\rho V c}{\mu} \qquad \Pi_3 \triangleq \frac{V}{a} \tag{3.2}$$

Π_1 incorporates dynamic pressure $\rho V^2/2$ and a reference area S (which is dimensionally equivalent to c^2). The other groups have simply been inverted. In their modified forms, Π_2 is the Reynolds Number[1] Re and Π_3 is the Mach number[2] MN. All of these quantities were introduced in Chapter 1. This enables Equation 3.1 to be re-expressed as:

$$F = \frac{1}{2}\rho V^2 S C_F(Re, MN, \alpha) \tag{3.3}$$

where F is the force and C_F is the associated force coefficient. A similar analysis can be performed for an aerodynamic moment M and this produces a moment coefficient C_M:

$$M = \frac{1}{2}\rho V^2 S c C_M(Re, MN, \alpha) \tag{3.4}$$

3.2.3 Lift, Drag, and Pitching Moment

Figure 3.5 shows a simple unswept wing in symmetric flight, where V is airspeed and α is AOA. The planform is defined by chord c and span b. Overall size and shape are defined by area S and aspect ratio A, respectively:

$$S = bc \qquad A = b/c \tag{3.5}$$

The summation of lift and drag distributions can be represented by vectors for lift L (perpendicular to the airstream) and drag D (parallel with the airstream), together with a pitching moment M. These are also drawn in Figure 3.5 at a reference centre located at the quarter chord ($c/4$). These quantities are expressed as follows:

$$L = \frac{1}{2}\rho V^2 S C_L \qquad D = \frac{1}{2}\rho V^2 S C_D \qquad M = \frac{1}{2}\rho V^2 S c C_M \tag{3.6}$$

where the dynamic pressure is $\rho V^2/2$. Quantities C_L, C_D, and C_M are coefficients of lift, drag, and pitching moment, respectively, which vary with AOA.

---

1 Reynolds Number gives the ratio of inertial forces to viscous forces in the boundary layer between a moving fluid and a bounding surface. It is used to predict the transition between laminar flow and turbulent flow.
2 Mach Number is the ratio of the speed of airflow and the speed of sound. It is used to predict the effects of compressibility in aerodynamic characteristics.

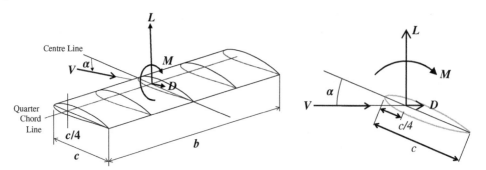

Figure 3.5 Isolated Wing in Symmetric Flight.

The lift coefficient can be written in two equivalent forms:

$$C_L = C_{L\alpha}(\alpha - \alpha_0) \tag{3.7a}$$

$$C_L = C_{L0} + C_{L\alpha}\alpha \tag{3.7b}$$

where $C_{L\alpha} = dC_{L\alpha}/d\alpha$ is the lift-curve gradient, α is the AOA, and α_0 is the zero-lift AOA. The equivalence is established as $C_{L0} = -C_{L\alpha}\alpha_0$. The product $C_{L\alpha}\alpha$ is called the induced lift (which is the lift induced by AOA). This is illustrated in Figure 3.6. It is noted that $C_{L0} = 0$ and $\alpha_0 = 0$ for a symmetric section. The linear characteristics apply in what are referred to as 'normal flight conditions', up to an AOA at which the lift gradient starts to reduce. Ultimately, the wing stalls and then the lift-curve gradient becomes negative.

The drag coefficient is written in the form:

$$C_D = C_{D0} + C_{Di} + C_{Dw} \tag{3.8}$$

where C_{D0} is the minimum drag coefficient (often assumed to be associated with zero-lift), C_{Di} is the induced drag coefficient, and C_{Dw} is the wave drag coefficient.

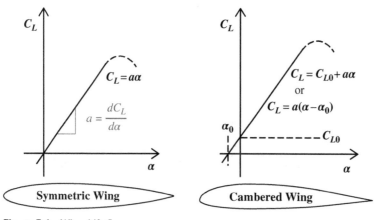

Figure 3.6 Wing Lift Curves.

The pitching moment coefficient is written in the form:

$$C_M = C_{M0} + C_{M\alpha}\alpha \tag{3.9}$$

The magnitude of the pitching moment depends on the point about which it is calculated. Invariably, the reference point is 25% MAC (as stated already). This is nominally coincident with the aerodynamic centre, which means that the pitching moment is constant. Therefore:

$$C_M = C_{M0} \tag{3.10}$$

3.2.4 Aerodynamic Centre

The *centre of lift* is the point on the wing chord at which the lift force L and the pitching moment M (defined at the moment reference centre) are replaced by an equivalent lift force (with zero pitching moment). This lies at distance x_{CP} from the quarter-chord, as shown in Figure 3.7. Taking clockwise moments as positive, the principle of equivalence gives:

$$x_{CP} = -\frac{M}{L} \tag{3.11}$$

The problem is that the centre of pressure becomes infinite as lift tends to zero. The solution is to exclude the zero-lift condition.

Lift and pitching moment vary linearly with AOA in normal flight:

$$L = L_0 + \frac{dL}{d\alpha}\alpha \qquad M = M_0 + \frac{dM}{d\alpha}\alpha$$

The lift force and pitching moment that are 'induced' by non-zero AOA can be replaced by an equivalent induced force at the so-called aerodynamic centre (or aero-centre). This is located at distance x_{AC} from the quarter-chord, as shown in Figure 3.8, where:

$$x_{AC} = -\frac{dM}{dL} \tag{3.12}$$

For modern wings, the aerodynamic centre lies close to the quarter-chord [$c/4$] for subsonic flow. It migrates to the half-chord position [$c/2$] for fully developed supersonic flow. The migration starts at the critical Mach number (at which the flow becomes locally supersonic somewhere on the wing).

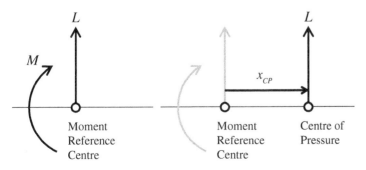

Figure 3.7 Centre of Pressure.

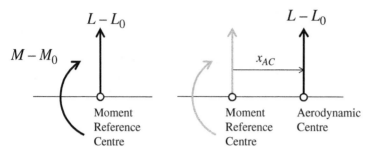

Figure 3.8 Aerodynamic Centre.

Figure 3.3 shows the thickness ratio of an aerofoil t/c, where t is the maximum thickness and c is the chord length. This is re-drawn in Figure 3.9, introducing the trailing edge angle ϕ_{TE}. These parameters have a small but significant effect of the position of the aerodynamic centre. This is shown in Figure 3.10.

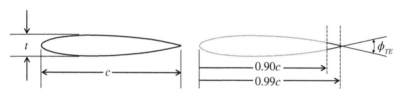

Figure 3.9 Aerofoil Cross-Section.

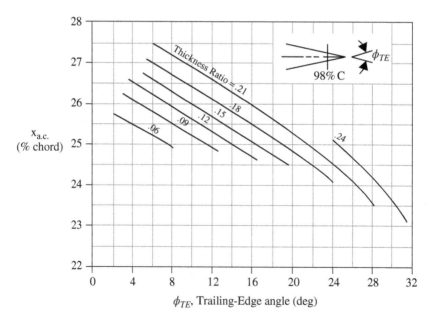

Figure 3.10 Effect of Aerofoil Cross-Sectional Geometry on the Aerodynamic Centre.

3.2.5 Wing Geometry

Figure 3.11 shows a generic, idealised swept wing. It has a centre-line chord length $[c_0]$, a tip chord length $[c_1]$, and span $[b]$. Sweep angles are denoted by the symbol Λ_n at the nth percentage point on the root chord. The chord length at distance y from the airfoil centre line is $c(y)$ and the local thickness ratio is written $t(y)/c(y)$. Wing geometric parameters are shown in Figure 3.12 and a formulary is given in Table 3.1. Note that the wing apex lies at the intersection of the leading edge and the centre line.

Aerodynamic loads (lift, drag, and pitching moment) are defined in Equation 3.6:

$$L = \frac{1}{2}\rho V^2 S\, C_L \qquad D = \frac{1}{2}\rho V^2 S\, C_D \qquad M = \frac{1}{2}\rho V^2 S\bar{c}\, C_M$$

where the dynamic pressure is $\rho V^2/2$. Quantities C_L, C_D and C_M are aerodynamic coefficients for lift, drag, and pitching moment, respectively, which vary with AOA. In addition, S is the wing area and $\bar{c}$ is the mean aerodynamic chord.

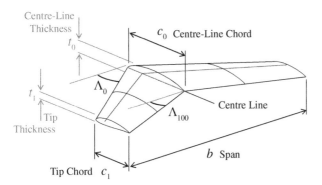

Figure 3.11 Idealised Swept Wing (Top Panel).

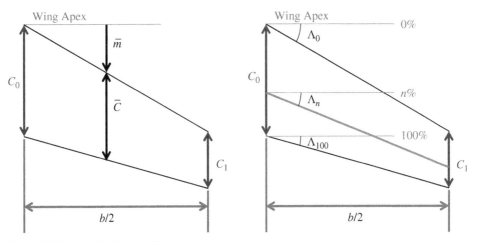

Figure 3.12 Idealised Swept Wing.

Table 3.1 Wing Geometric Parameters.

Taper ratio	$\lambda = c_1/c_0$	Mean geometric chord	$c' = \dfrac{1+\lambda}{2}c_0$
Aspect ratio	$A = b/c'$	Mean aerodynamic chord	$\bar{c} = \left(\dfrac{2}{3}\right)\dfrac{1+\lambda+\lambda^2}{1+\lambda}c_0$
Planform area	$S = bc'$	Leading edge of mean aerodynamic chord	$\bar{m} = \left(\dfrac{b}{6}\right)\dfrac{1+2\lambda}{1+\lambda}\tan\Lambda_0$
Thickness ratio	$\tau = \dfrac{t}{c}$	Wetted area	$S_{wet} = 2S\left(1+\dfrac{1}{4}\tau_0\dfrac{1+\lambda\tau_1/\tau_0}{1+\lambda}\right)$
Sweep angle of $n\%$ chord line		$\tan\Lambda_n = \tan\Lambda_0 - \dfrac{n}{100}\left(\dfrac{4}{A}\right)\dfrac{1-\lambda}{1+\lambda}$	

The use of coefficients provides a compact way of expressing the total aerodynamic loads. They are obtained by integrating the load distribution across the wing span. Thus, Equation 3.10 can be rewritten as:

$$L = \frac{1}{2}\rho V^2 \int_{-b/2}^{b/2} c(y)C_L(y)dy \quad D = \frac{1}{2}\rho V^2 \int_{-b/2}^{b/2} c(y)C_D(y)dy \quad M = \frac{1}{2}\rho V^2 \int_{-b/2}^{b/2} c^2(y)C_M(y)dy$$

Hypothetically, if the coefficients are constant across the span, then this can be simplified as follows:

$$L = \frac{1}{2}\rho V^2 C_L \int_{-b/2}^{b/2} c(y)dy \quad D = \frac{1}{2}\rho V^2 C_D \int_{-b/2}^{b/2} c(y)dy \quad M = \frac{1}{2}\rho V^2 C_M \int_{-b/2}^{b/2} c^2(y)dy$$

Therefore, wing area and mean aerodynamic chord are defined as:

$$S = \int_{-b/2}^{b/2} c(y)dy \qquad \bar{c} = \frac{1}{S} \int_{-b/2}^{b/2} c^2(y)dy \qquad (3.13)$$

In contrast, the mean geometric chord is defined as:

$$c' = \frac{1}{b} \int_{-b/2}^{b/2} c(y)dy \qquad (3.14)$$

Note that the aerodynamic centre is defined at 25% of the mean aerodynamic chord.

3.2.6 NACA Four-Digit Sections

As an example of cross-sectional geometry, a NACA four-digit aerofoil[3] is defined in Figure 3.13. Sections are designated using four digits with the prefix 'NACA'. All lengths

---

3 Source: NACA-RP-460.

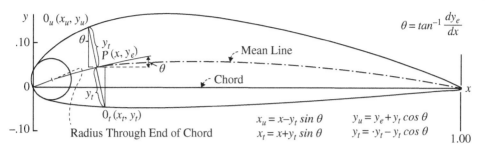

Figure 3.13 NACA Four-Digit Section Geometry.

are normalized with respect to a unit chord length. The first digit gives the maximum camber [m] in percent. The second digit gives the position of the maximum camber [p] in tenths. The last two digits give the maximum thickness [t] in percent. Thus, for example, NACA0012 aerofoil is symmetric with thickness ratio of 12%. NACA2310 has maximum camber of 2% at 30% chord and thickness ratio of 10%.

The camber line is defined in two parts, relative to the point of maximum camber [$x = m$]. Forward of that point ($0 \leq x \leq m$):

$$y_c = \frac{m}{p^2}\left(2px - x^2\right)$$

Aft of that point ($m \leq x \leq 1$):

$$y_c = \frac{m}{(1-p)^2}\left((1 - 2p) + 2px - x^2\right)$$

The gradient of the camber line at any value of x is:

$$\tan\theta = \frac{dy_c}{dx}$$

The thickness profile with respect to the camber line is given by:

$$y_t = \frac{t}{0.2}\left(0.29690\sqrt{x} - 0.12600x - 0.35160x^2 + 0.28430x^3 - 0.10150x^4\right)$$

Noting that the geometry is based on a curvilinear coordinate system, the XY-coordinates of the upper and lower surfaces (with subscripts 'u' and 'l', respectively) are defined as follows:

$$x_u = x - y_t \sin\theta \qquad y_u = y_c + y_t \cos\theta$$
$$x_l = x + y_t \sin\theta \qquad y_l = y_c - y_t \cos\theta$$

Finally, the leading-edge radius (which is annotated as 'Radius through end of chord') is:

$$r_{LE} = \frac{1}{2}\left(\frac{0.29690}{0.20}t\right)^2 = 0.1019t^2$$

For information, Table 3.2 presents data for a selection of NACA four-digit profiles, namely 2D lift-curve gradient [a_{2D}], zero-lift AOA [α_0], zero-lift pitching moment [C_{M0}],

Table 3.2 Aerodynamic Properties for NACA Four-Digit Sections.

NACA Section	a_0	α_0	C_{M0}	C_{D0}	C_{Li}	n
NACA0012	0.101	0	−0.002	0.0083	0	0.9
NACA2312	0.102	−1.7	−0.036	0.0088	0.10	0.3
NACA2412	0.103	−1.7	−0.042	0.0090	0.15	0.4
NACA2512	0.102	−2.0	−0.050	0.0089	0.15	0.3
NACA4312	0.102	−3.7	−0.072	0.0101	0.23	0.5
NACA4412	0.100	−3.9	−0.087	0.0085	0.33	0.5
NACA4512	0.101	−4.0	−0.102	0.0103	0.27	0.8
NACA6312	0.102	−5.4	−0.108	0.0108	0.40	0.4
NACA6412	0.102	−5.7	−0.129	0.0104	0.42	0.6
NACA6512	0.101	−6.3	−0.154	0.0101	0.42	0.7

minimum drag coefficient [C_{D0}], and ideal lift coefficient [C_{Li}] (corresponding with minimum drag). Also, it defines the position of the aerodynamic centre ahead of the quarter chord [n] according to the relationship:

$$C_{M\left(\frac{c}{4}\right)} = C_{M0} + \frac{n}{100} C_L$$

3.3 Aerodynamic Model of an Isolated Wing

3.3.1 Aerodynamic Lift

Aerodynamic lift is calculated according to Equation 3.6:

$$L = QS\, C_L$$

where $Q = \rho V^2/2$ is the dynamic pressure, S is the panform area, and C_L is the lift coefficient. Under normal flight conditions, the lift coefficient is given by Equation 3.7:

$$C_L = C_{L0} + a\alpha \tag{3.15}$$

where $a = dC_L/d\alpha$ is the lift-curve gradient. This can be approximated using DATCOM[4]:

$$a = \frac{2\pi A}{2 + \sqrt{4 + (A/\kappa)^2 K_M^2 (1 + \tan \Lambda_M)^2}} \tag{3.16}$$

---

4 Source: US Airforce Stability and Control Data Compendium (DATCOM).

where

$$K_M = \sqrt{1 - M_N^2} \qquad \text{and} \qquad \tan \Lambda_M = \frac{\tan \Lambda_{50}}{K_M}$$

and where M_N is the flight Mach number, A is the aspect ratio, and Λ_{50} is the mid-chord sweep angle.

For a wing for which $M_N \longrightarrow 0$, $A \longrightarrow \infty$, and $\Lambda_{50} \longrightarrow 0$, this becomes:

$$a \longrightarrow a_{2D} = 2\pi\kappa \text{ where } k = \frac{a_{2D}}{2\pi} \tag{3.17}$$

For a hypothetical wing of infinite span, the local flow pattern is aligned with the local wing chord. The flow streamlines have a two-dimensional (2D) pattern that is repeated everywhere. A real wing has finite span and the flow over upper and lower surfaces has to mix in order to equalise the pressure at the wing tips. This is achieved by flow under the wing being drawn around the tips by the lower pressure above the wing. For the outer wing, the flow direction turns outwards over the lower surface and inwards over the upper surface. This exhibits a three-dimensional (3D) flow pattern.

Thus, parameter a [in Equation 3.16] is the 3D lift-curve gradient and parameter a_{2D} is the 2D lift-curve gradient at low speed. The theoretical lift-curve gradient can be approximated as:

$$a_{theory} \approx 2\pi + 4.7\,\tau\,(1 + \tan \phi_{TE}) \tag{3.18}$$

where τ is the thickness ratio and ϕ_{TE} is the trailing-edge angle shown in Figure 3.9. An alternative approximation is often used:

$$a_{theory} \approx 2\pi + 4.9\,\tau \tag{3.19}$$

Either way, $a_{theory} \longrightarrow 2\pi$ for an ideal thin wing. This is expressed as the ratio that appeared in Equation 3.16:

$$\kappa = \frac{a_{2D}}{2\pi} \longrightarrow 1 \tag{3.20}$$

The actual lift-curve slope is calculated as follows:

$$a_{2D} = 1.05\,K_a\,a_{theory} \tag{3.21}$$

where K_a is an empirical correction (shown in Figure 3.14) that is a function of Reynolds Number Re and trailing-edge angle ϕ_{TE}. In addition, an example is shown for a typical wing section over the range of Reynolds Number from 10^6 to 10^8.

The Reynolds Number Re gives the ratio of inertial forces to viscous forces in the boundary layer between a moving fluid and a bounding surface. It is used to predict the transition between laminar flow and turbulent flow. It is defined as $Re = \rho Vl/\mu$, where ρ is air density, μ is dynamic viscosity, V is the flight velocity, and l is the reference length for the particular component being considered (which is the wing chord length in this context). Dynamic viscosity is calculated using Sutherland's formula [as in Volume 1] such that $\mu = \beta\sqrt{T^3}/(T + S)$, where $S = 110$ K and $\beta = 1.458 \times 10^{-6}$ N. s. m^{-2}. K$^{-0.5}$.

Equation 3.7b can now be rewritten as:

$$C_L = a_{3D}\,(\alpha - \alpha_0) \tag{3.22}$$

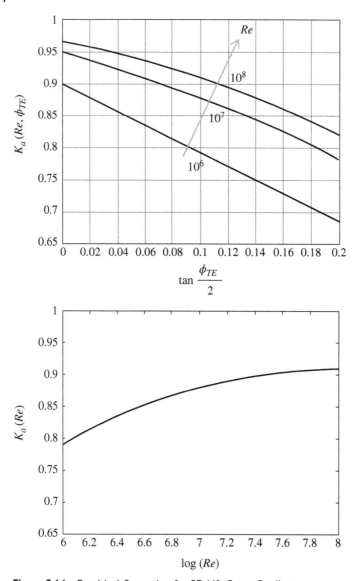

Figure 3.14 Empirical Correction for 2D Lift-Curve Gradient.

If the zero-lift AOA [$\alpha_{0,2D}$] is known for a 2D section (e.g. from Section 3.2.6), then the corresponding 3D value is given by:

$$\tan \alpha_0 = \frac{\tan \alpha_{0,2D}}{\cos \Lambda_{25}} \tag{3.23}$$

3.3.2 Pitching Moment

Aerodynamic pitching moment is calculated according to Equation 3.10:

$$M = QSc\, C_M$$

where $Q = \rho V^2/2$ is the dynamic pressure, S is the panform area, c is the mean aerodynamic chord, and C_M is the pitching moment coefficient. Given that pitching moment is calculated at the aerodynamic centre, Equation 3.14 dictates that $C_M = C_{M0}$. For a 2D wing section, if the zero-lift pitching moment coefficient $[C_{M0,2D}]$ is known (e.g. from Section 3.2.6), then the corresponding value for the 3D wing is given by:

$$C_{M0} = \frac{A \cos^2 \Lambda_{25}}{A + 2\cos \Lambda_{25}} C_{M0,2D} \qquad (3.24)$$

where $A =$ aspect ratio and $\Lambda_{25} =$ quarter-chord sweep angle (which is calculated from Table 3.1).

3.3.3 Drag Force

Aerodynamic drag is calculated according to Equation 3.2:

$$D = QS\,C_D$$

where $Q = \rho V^2/2$ is the dynamic pressure, S is the panform area, and C_D is the drag coefficient. In turn, this is defined by Equation 3.5:

$$C_D = C_{D0} + C_{Di} + C_{Dw}$$

where the coefficients are C_{D0} (profile drag), C_{Di} (induced drag), and C_{Dw} (wave drag). Drag estimation is the subject of numerous publications and, here, it is convenient to refer selectively to literature reviews [such as Takahashi et al. (2010) and Zold (2012)].

3.3.4 Profile Drag

Profile drag is caused by air having to force its way around the wing as a physical obstruction. Its coefficient can be expressed as:

$$C_{D0} = C_f F \frac{S_{wet}}{S_{ref}} \qquad (3.25)$$

where C_f is the flat-plate skin friction coefficient, F is the form factor, and S_{wet}/S_{ref} is the ratio of wetted surface area to the reference planform area. The wetted area is defined in Table 3.1.

The flat-plate skin friction coefficient is defined for a combination of laminar and turbulent flow:

$$C_f = \lambda\, C_{f,laminar} + (1-\lambda)\, C_{f,turbulent} \qquad (3.26)$$

where λ is fraction of the chord that has laminar flow [where it is assumed that $\lambda \approx 0.3$ for a typical wing and $\lambda \approx 0.2$ for a typical tail surface]. The laminar and turbulent components are:

$$C_{f,laminar} = \frac{1.328}{\sqrt{Re}} \qquad C_{f,turbulent} = \frac{0.455}{(log\,Re)^{2.58}\left(1 + 0.144M_N^2\right)^{0.65}} \qquad (3.27)$$

where M_N is Mach number and Re is Reynolds Number. When estimating skin friction for a rough surface, problems are avoided by limiting the value of Reynolds Number. For subsonic aerodynamics:

$$Re \leq 38.21 \left(\frac{c}{k}\right)^{1.053} \tag{3.28}$$

where c is the chord length and k is the skin roughness factor (0.634×10^{-5} m for smooth paint).

The form factor for a wing is given by:

$$F = 1.34 \left[1 + 0.6 \left(\frac{c}{x}\right)\tau + 100\tau^2\right] M^{0.18}(\cos \Lambda_m)^{0.28} \tag{3.29}$$

where x/c gives the location of the maximum thickness ratio τ as a fraction of the wing chord (typically 0.3–0.5). The angle Λ_m is the sweep angle at the line of maximum thickness.

By way of historical context, Figure 3.15 summarises the results of full-scale wind tunnel tests on a Grumman Avenger published in Lange (1945). The interesting and surprising conclusion is the large increase in C_{D0} from the basic airframe to the service-ready configuration (due to the many modifications required for aircraft at that time).

3.3.5 Induced Drag

The easiest interpretation of induced drag is based on a wing (with span b) flying at constant speed V through a flow tube[5] with cross-sectional area $\pi(b/2)^2$. When air flows over a simple wing at a positive AOA, lift is generated and (as a by-product) the airstream direction is deflected downwards when it leaves the trailing edge. This is defined by a downwash angle ε, as shown in Figure 3.16. The mean airstream direction over the wing is the *induced* AOA $\alpha_i \approx \varepsilon/2$. In this context, the lift force must be consistent with a momentum force in the vertical direction, such that:

$$L = \frac{1}{2}\rho V^2 S\, C_L \equiv \left[\rho \pi \left(\frac{b}{2}\right)^2 V\right] V \sin \varepsilon$$

The term in brackets defines a constant massflow through the flow tube. Assuming small angles, together with the induced AOA, this equation becomes:

$$\frac{1}{2}\rho V^2 S\, C_L \approx \left[\rho \pi \left(\frac{b}{2}\right)^2 V\right] V(2\alpha_i) = \frac{1}{2}\rho V^2 \left(\pi b^2\right)\alpha_i$$

Therefore, the induced AOA is determined as follows:

$$\alpha_i \approx \frac{C_L}{\pi A} \tag{3.30}$$

---

5 A typical undergraduate question is 'why does the flow tube have to have a circular cross-section?'. The lecturer responds by saying 'Good question!'.

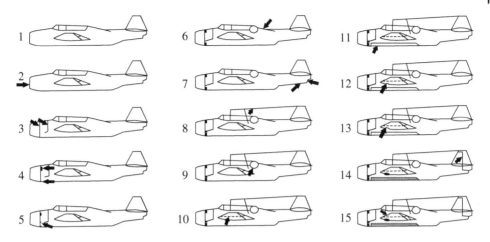

	Configuration	C_{D0}
1	Airplane Completely Sealed and Faired	**0.0183**
2	Flat Plate Removed from Nose	**0.0189**
3	Seals Removed from Flapped-cowling Air Exits	**0.0199**
4	Seals Removed from Cowling-flap Hinge-line Gaps	**0.0203**
5	Exhaust Stacks Replaced	**0.0211**
6	Canopy Fairing Removed, Turret Leaks Sealed	**0.0222**
7	Tail Wheel and Arresting-hook Opening Uncovered	**0.0223**
8	Aerial, Mast and Trailing Antenna Tube Installed	**0.0227**
9	Canopy and Turret Leak Seals Removed	**0.0230**
10	Leak Seals Removed from Shock Strut, Cover Plate and Wing-fold Axis	**0.0234**
11	Leak Seals Removed from Bomb-bay Doors and Miscellaneous Seals Removed	**0.0236**
12	Fairings Over Catapult Hooks Removed	**0.0237**
13	Wheel-well Cover Plates Removed	**0.0251**
14	Seals Removed from Tail-surface Gaps	**0.0260**
15	Plates Over Wing-tip Slot Openings Removed. Airplane in Service Condition.	**0.0264**

C_{D0} measured at $C_L = 0.245$ 44% Increase

Figure 3.15 C_{D0} Measurements for a Grumman Avenger.

where the aspect ratio A is obtained from Table 3.1. The induced AOA can be visualised as the angle between vectors L and R in Figure 3.16. Assuming small angles, the following approximations apply:

$$L = R \cos \alpha_i \approx R \qquad D = R \sin \alpha_i \approx R \alpha_i \tag{3.31}$$

The prediction of induced drag is obtained as:

$$D \approx L \alpha_i \approx L \frac{C_L}{\pi A} \tag{3.32}$$

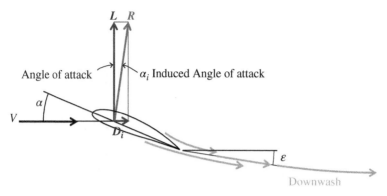

Figure 3.16 Conceptual Model of Airflow over a Wing.

Inevitably drag is always greater than hoped and this is quantified using an efficiency factor e. This gives the standard expression for induced drag (in coefficient form):

$$C_{Di} = \frac{C_L^2}{\pi A e} \tag{3.33}$$

For initial calculations, wing efficiency might be estimated as $e \approx 2.5/\pi$, i.e. $(1/\pi e) \approx 0.4$. Apart from this, there are many different formulas for calculating wing efficiency, such as Raymer (1999) and Pamadi (2015).

Niţă and Scholz (2012) propose the following method for calculating wing efficiency. They note the coupling between taper ratio and sweep angle that results in an almost elliptical lift distribution, which is known to minimise the induced drag. NACA-RP-921 [DeYoung and Harper 1955] propose a coupling relationship that can be approximated as follows:

$$\lambda_{opt} = 0.45\varepsilon^{-0.0375\Lambda_{25}} \tag{3.34}$$

where λ_{opt} = optimum taper ratio, Λ_{25} = quarter-chord sweep angle, and ε = Napier's constant.[6]

The theoretical efficiency (without corrections) is drawn from Hörner (1951):

$$e_{theory} = \frac{1}{1 + f(\lambda)A} \tag{3.35}$$

where A = aspect ratio and the function of taper ratio is approximated as:

$$f(\lambda) \approx 0.0524\lambda^4 - 0.15\lambda^3 + 0.1659\lambda^2 - 0.0706\lambda + 0.0119 \tag{3.36}$$

The optimum efficiency is achieved when the value of $f(\lambda)$ is minimized, i.e. when $\lambda_{opt} = 0.356590$. However, this is only valid for unswept wings. For swept wings, Equation 3.27 is adjusted in order to align the optimum efficiency with the prediction from Equation 3.26:

$$e_{theory} = \frac{1}{1 + f(\lambda - \Delta\lambda)A} \tag{3.37}$$

---

6 Base of natural logarithms: $\varepsilon = 2.718281828459046$.

where

$$\Delta\lambda = \left(0.45\varepsilon^{-0.0375\Lambda_{25}}\right) - 0.356590 \tag{3.38}$$

The overall efficiency including corrections is given by:

$$e = \frac{K_F K_{D0} K_M}{1 + 0.0118A} \tag{3.39}$$

For jet aircraft, the fuselage correction is $K_F \approx 0.97$ and zero-lift drag correction is $K_{D0} \approx 0.87$. The Mach number correction is:

$$K_M = 1 - 0.001521 \left(\frac{max(M,0.3)}{0.3} - 1\right)^{10.82} \tag{3.40}$$

3.3.6 Wave Drag

Wave drag is due to the progressive development of local regions of supersonic flow and the formation of shock waves. This process commences above the *Critical Mach Number*, which is the maximum speed at which the flow field across the entire wing is subsonic. The variation of drag with Mach number has been established exhaustively via experimental observations. Mason (1990) and Sforza (2014) both have offered a succinct account of various approaches to this phenomenon.

Drag divergence is subject to varying interpretations and empirical estimates. Lock's approximation for wave drag is given by:

$$C_{Dw} = 20 \left(M - M_{crit}\right)^4 \tag{3.41}$$

and the McDonnell–Douglas criterion associated the drag divergence Mach number M_{dd} with the condition $dC_{Dw}/dM = 0.1$. Thus, based on Equation 3.33:

$$M_{dd} = M_{crit} + 0.1078 \tag{3.42}$$

Korn's Equation gives a good initial estimate for M_{dd} in the form:

$$M_{dd} = \frac{k_T}{\cos\Lambda} - \frac{\tau_{max}}{\cos^2\Lambda} - \frac{kC_L}{\cos^3\Lambda} \tag{3.43}$$

where τ_{max} is the maximum thickness ratio, C_L is the lift coefficient where Λ is the sweep angle (as used by Mason), which is stated elsewhere (by Sforza) as the half-chord sweep angle. The parameter k_T is called a 'technology factor' and is usually in the range 0.87–0.955 and k is a slope factor that is usually in the range 0.1–0.14 [Wislicenus and Daidzic (2022), Filippone (2012)].

A DATCOM method exists for determining the drag divergence Mach number. This is reproduced in Figure 3.17 and applied to rectangular wings with different aspect ratio and thickness/chord ratios. In the general case, this also has dependencies on the quarter-chord sweep angle, the taper ratio, and the 'general configuration' of a wing.

The significance of drag divergence is wrapped up in transonic aerodynamics, which is not covered in this text. However, it is worth being aware of how parameters vary as Mach number increases. Figure 3.18 shows the variations in C_L and C_M (taken from the author's

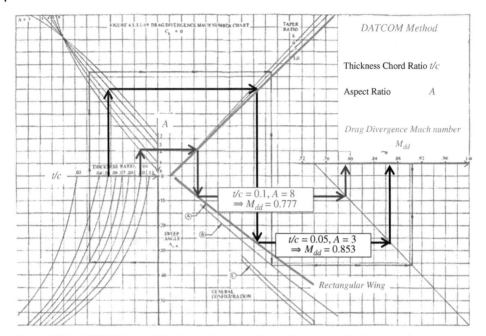

Figure 3.17 DATCOM Method for Determination of Drag Divergence Mach number.

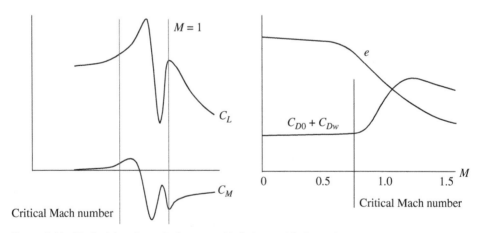

Figure 3.18 Typical Aerodynamic Parameter Variation vs. Mach number.

engineering scrapbook) and the variations in wing efficiency e and the combination of profile drag and wave drag $C_{D0} + C_{Dw}$. Note that these are sketches; the actual variations will be highly dependent on the aircraft configuration. Figure 3.19 show the quantitative variation in wing profile drag (or minimum drag) against thickness ratio and sweep angle.

Sweep angle measured at quarter-chord line

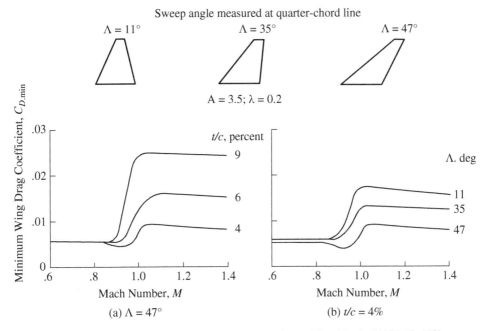

$\Lambda = 11°$ $\Lambda = 35°$ $\Lambda = 47°$

$A = 3.5; \lambda = 0.2$

Figure 3.19 Wing Profile Drag vs Sweep Angle and Thickness/Chord Ratio [NASA-SP-468].

3.4 Trailing-Edge Controls

Trailing-edge flaps were introduced in Chapter 1 and are shown again in Figure 3.20. These can be full-span (e.g. an *elevator* on an aircraft tail surface) or part-span (e.g. *flaps* and *ailerons* on an aircraft wing). Symmetric deployment generates lift and anti-symmetric deployment generates a rolling moment. In principle, deployment could be arbitrarily asymmetric, in which case the controls would be called *flaperons*.

3.4.1 Incremental Lift

The incremental change in 2D lift coefficient due to flap rotation δ is:

$$\Delta C_{L,2D} = C_{L\delta,2D}\,\delta = C_{L\delta,theory}\,K_\delta \delta \tag{3.44}$$

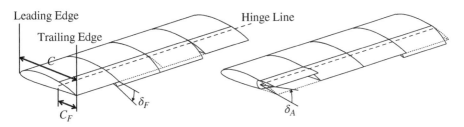

Figure 3.20 Trailing-Edge Control Surfaces.

Table 3.3 Numerical Approximations for Two-Dimensional Flap Effects.

P	w	Approximation: $P = wP_1 + (1-w)P_2$
$C_{L\delta,theory}$	$\dfrac{\left(\frac{t}{c}\right)}{0.15}$	$P_1 = -9.888889 \left(\frac{c_F}{c}\right)^2 + 14.661111 \left(\frac{c_F}{c}\right) + 0.311111$ $P_2 = -10.333333 \left(\frac{c_F}{c}\right)^2 + 13.016667 \left(\frac{c_F}{c}\right) + 1.175000$
K_δ	$\dfrac{1 - \left(\frac{a_{2D}}{a_{theory}}\right)}{1 - 0.7}$	$P_1 = 0.777778 \left(\frac{c_F}{c}\right) + 0.311111$ $P_2 = 1$

This introduces a lift-curve slope due to flap rotation $C_{L\delta,2D}$; also known as *lift effective-ness*. For 2D flow, there is a theoretical value $C_{L\delta,theory}$ and an empirical correction K_δ, which are approximated in Table 3.3. The relationship between 2D and 3D lift increments is as follows:

$$\Delta C_L = (K_b(\eta_{ob}) - K_b(\eta_{ib})) K_c K_F \frac{a}{a_{2D}} \Delta C_{L,2D} \tag{3.45}$$

Key parameters are the flap chord factor K_c, flap span factor K_b, and an empirical correc-tion K_F. The flap span is defined by inboard and outboard edges, η_{ib} and η_{ob}, where η is the nondimensional spanwise position in the range $[-1, 1]$. The 2D and 3D lift-curve slopes are a_{2D} and a_{3D}, respectively.

The flap span factor and the empirical correction are approximated in Table 3.4 and depicted in Figures 3.21 and 3.22. The flap chord factor is approximated by hyperbolic curves as shown in Figure 3.23. These are calculated as:

$$K_c = 1 + \frac{k}{0.07 + 0.1A} \tag{3.46}$$

where A = aspect ratio and k = hyperbolic constant. The hyperbolic constant is $k = r^2/2$, where r = reference 'distance' of the hyperbolic curve from its geometric origin (which is not the origin of coordinates). This is defined by:

$$r = \frac{1}{6.6}\left(3.6 - 3\alpha_\delta + 9p_\delta^2\right) \qquad \text{where} \qquad p_\delta = \max(0, 0.4 - \alpha_\delta) \tag{3.47}$$

Table 3.4 Numerical Approximations for Three-Dimensional Flap Effects.

P	w	Approximation: $P = wP_1 + (1-w)P_2$
K_b	λ Taper ratio	$P_1 = -0.4\,\eta^2 + 1.4\,\eta$ $P_2 = -0.7\,\eta^2 + 1.7\,\eta$
K_F	$\dfrac{0.5 - \left(\frac{c_F}{c}\right)}{0.1 - 0.1}$	$P_1 = v_1 + (1 - v_1)\,0.9(0.80 - 0.004\,\delta)$ $P_1 = v_1 + (1 - v_1)\,0.9(0.66 - 0.004\,\delta)$ where $v_1 = \dfrac{1}{2}\left(1 - \tanh\dfrac{\delta - 21}{5}\right) \qquad v_2 = \dfrac{1}{2}\left(1 - \tanh\dfrac{\delta - 18}{4}\right)$

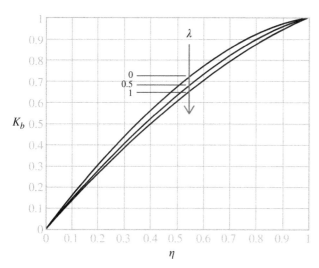

Figure 3.21 Flap Span Factor.

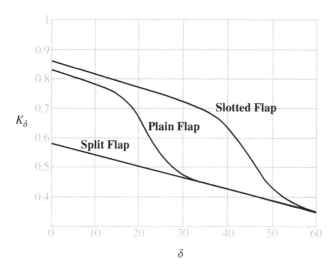

Figure 3.22 Correction Factor for Flaps.

The spacing between curves is almost uniform for larger values of α_δ. The additional (non-linear) term is needed because the spacing increases as α_δ decreases.

The parameter α_δ is the flap effectiveness, which is the equivalent change in AOA per unit of flap deflection. This is shown in Figure 3.24, calculated using the formula:

$$\alpha_\delta = 1 - \frac{\varphi - \sin\varphi}{\pi} \qquad \text{where} \qquad \cos\varphi = 2\frac{c_F}{c} - 1 \tag{3.48}$$

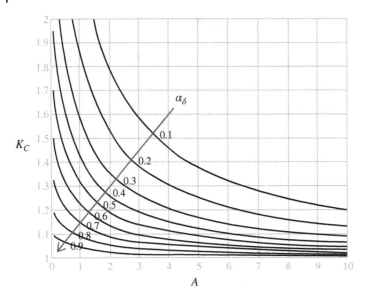

Figure 3.23 Flap Chord Factor.

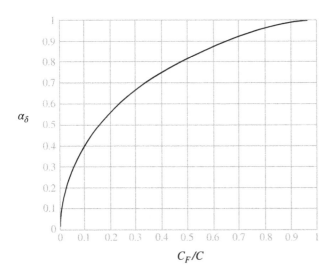

Figure 3.24 Flap Effectiveness.

3.4.2 Incremental Drag

The incremental drag coefficient associated with deflection of a plain flap is estimated as:

$$\Delta C_{D0} = 1.7 \left(\frac{c_F}{c}\right)^{1.38} \left(\frac{S_F}{S}\right) \sin^2 \delta \tag{3.49}$$

where c_F/c is the flap-chord ratio and S_F/S is flap area ratio (i.e. ratio of areas taken across the flap span and full span). Note that it is common to account for simultaneous flap deflection on both sides of the wing centre line.

3.4.3 Incremental Pitching Moment

Flap deflections cause incremental changes to local pitching moment, which is equivalent to a movement of the local aerodynamic centre. The chordwise lift distribution increases when a flap is deployed and there is a peak in the vicinity of the hinge (where surface curvature is greatest). The overall effect is more lift, as discussed already, and a nose-down pitching moment. The aerodynamic centre moves aft. Given a change in local lift coefficient $[\Delta C_l]$, the corresponding change in local pitching moment coefficient $[\Delta C_m]$ is a function of flap chord ratio $[c_F/c]$:

$$\Delta C_m = -\frac{0.1}{0.325}\left(0.805 - \frac{c_F}{c}\right)\Delta C_l \tag{3.50}$$

Recalling Section 4.3.1.6, the shift in local aerodynamic centre is:

$$\Delta x_{AC} = -c\,\frac{\Delta C_m}{\Delta C_l} = 0.2476 - 0.3076\,\frac{c_F}{c} \tag{3.51}$$

3.4.4 Hinge Moments

The aerodynamic loading on flaps is crucial in determining the control forces. These are applied to some form of mechanical linkage that is attached to the flap in order to rotate it to a given position at a given rate. For manual controls, it is very important to provide a system that can be driven by a human pilot. For augmented controls (which are driven by actuators), it is very important to a system that does not cause flaps to be fail structurally and become detached from the airframe. To this end, empirical relationship will be stated for the calculation of hinge moments.

The hinge moment on a flap can be written in the form:

$$H = QS_F c_F(h_0 + h_\alpha\alpha + h_\delta\delta) \tag{3.52}$$

where $Q = \rho V^2/2$ is the dynamic pressure, $S_F =$ flap area, $c_F =$ flap chord length, $\alpha =$ AOA, and $\delta =$ flap deflection. In the absence of data on airfoil cross-sections, preliminary calculations can proceed using $h_0 = 0$.

Aerodynamic balancing is used to resolve conflicting requirements for small hinge moments (for ease of operation) and large hinge moments (for disturbance rejection). Without discussing in detail, it is noted that this is heavily influenced by the leading-edge profile of the flap. This is quantified as the *balance ratio*:

$$BR = \sqrt{\left(\frac{c_B}{c_F}\right)^2 + \left(\frac{t_B}{2c_F}\right)^2} \tag{3.53}$$

where c_F is the flap chord length, c_B is the distance between the flap leading-edge and the hinge line, and t_B is the flap thickness at the hinge line. For round-nose flaps, $c_B = t_B/2$ and $BR = \sqrt{2}(c_B/c_F)$.

The hinge moment derivatives are established with *balanced* and *incremental* components:

$$h_\alpha = w_\alpha h_{\alpha,bal} + \Delta h_\alpha \tag{3.54}$$

$$h_\delta = (h_{\delta,bal} - w_\delta h_{\alpha,bal}\alpha_\delta + \Delta h_\delta)\cos\Lambda_{25}\cos\Lambda_{HL} \tag{3.55}$$

where Λ_{25} is the quarter-chord sweep angle and Λ_{HL} is the hinge-line sweep angle. The term α_δ is the flap effectiveness, as defined in Equation 4.38. Recall that this is the equivalent change in AOA per unit of flap deflection.

There are two related weighting factors, based on aspect ratio and quarter-chord sweep angle:

$$w_\alpha = \frac{A\cos\Lambda_{25}}{2 + A\cos\Lambda_{25}} \qquad w_\delta = \frac{2\cos\Lambda_{25}}{2 + A\cos\Lambda_{25}} \tag{3.56}$$

The balanced derivatives are:

$$h_{\alpha,bal} = B_\alpha\left(\frac{h_{H\alpha}}{h_{\alpha,theory}}\right)h_{\alpha,theory} \qquad h_{\delta,bal} = B_\delta\left(\frac{h_{H\delta}}{h_{\delta,theory}}\right)h_{\delta,theory} \tag{3.57}$$

The incremental derivatives are:

$$\Delta h_\alpha = B_2 K_\alpha\left(\frac{K_{H\alpha}(\eta_{OB}) - K_{H\alpha}(\eta_{IB})}{\eta_{OB} - \eta_{IB}}\right)C_{l\alpha}\cos\Lambda_{25} \tag{3.58a}$$

$$\Delta h_\delta = B_2 K_\delta\left(\frac{K_{H\delta}(\eta_{OB}) - K_{H\delta}(\eta_{IB})}{\eta_{OB} - \eta_{IB}}\right)C_{l\delta}\cos\Lambda_{25} \tag{3.58b}$$

Numerical approximations are specified in Tables 3.5 and 3.6.

Table 3.5 Further Numerical Approximations for Flap Hinge Moments.

P	Approximation: $P = a_2 X^2 + a_1 X + a_0$	P	Approximation: $P = a_1 X + a_0$
$K_{H\alpha}$	$3.0\,\eta^2 + 0.2\,\eta + 1$	B_α	$-2.514286\,BR + 1.277143$
$K_{H\delta}$	$3.6\,\eta^2 + 0.3\,\eta + 1$	B_δ	$-0.933333\,BR + 0.966667$
K_α	$0.01125\,A^2 - 0.11375\,A + 0.36250$		

Table 3.6 Numerical Approximations for Flap Hinge Moments.

P	w	Approximation: $P = wP_1 + (1-w)P_2$	
$h_{\alpha,theory}$	$\dfrac{t/c}{0.15}$	$P_1 = -1.40\left(\dfrac{c_F}{c}\right) - 0.09$	$P_2 = -1.366667\left(\dfrac{c_F}{c}\right) - 0.203333$
$\dfrac{h_{H\alpha}}{h_{\alpha,theory}}$	$\dfrac{1-\left(a_{2D}/a_{2D,theory}\right)}{1-0.7}$	$P_1 = 0.80\left(\dfrac{c_F}{c}\right) - 0.18$	$P_2 = 1$
$h_{\delta,theory}$	$\dfrac{t/c}{0.15}$	$P_1 = 1.40\left(\dfrac{c_F}{c}\right) + 0.50$	$P_2 = -1.60\left(\dfrac{c_F}{c}\right) + 0.72$
$\dfrac{h_{H\delta}}{h_{\delta,theory}}$	$\dfrac{1-\left(a_{2D}/a_{2D,theory}\right)}{1-0.6}$	$P_1 = -0.60\left(\dfrac{c_F}{c}\right) + 0.70$	$P_2 = 1$
B_2	$\dfrac{0.3-\left(c_F/c\right)}{0.3-0.1}$	$P_1 = -1.142857\left(\dfrac{c_F}{c}\right)^2 + 0.034286\left(\dfrac{c_F}{c}\right) + 0.490000$	
		$P_2 = -1.714286\left(\dfrac{c_F}{c}\right)^2 - 0.134286\left(\dfrac{c_F}{c}\right) + 1.098571$	
K_δ	$\dfrac{\left(c_F/c\right)-0.2}{0.6-0.2}$	$P_1 = 0.000550A^2 - 0.009850A + 0.047500$	
		$P_2 = 0.000858A^2 - 0.014675A + 0.065917$	

3.5 Factors affecting Lift Generation

3.5.1 Sideslip

Sideslip is the lateral component of the velocity vector. It is zero in symmetric flight. It mainly occurs when flying through a cross-wind or flying with asymmetric thrust. Figure 3.25 shows the standard definitions of AOA and Angle of Sideslip (AOS), based on velocities (u, v, w). Note that AOA is defined in the longitudinal plane of symmetry and AOS accounts for cross-flow:

$$\tan \alpha = \frac{w}{u} \qquad \sin \beta = \frac{v}{V} \tag{3.59}$$

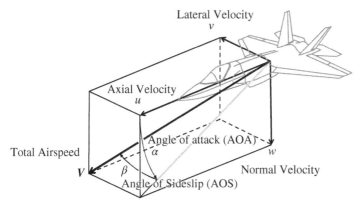

Figure 3.25 Airflow Direction.

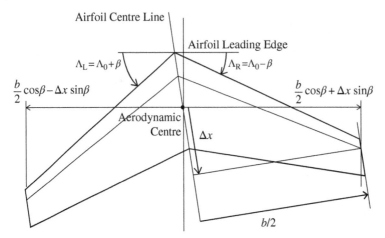

Figure 3.26 Wing Geometry in Sideslip.

Figure 3.26 shows a wing that is flying pure sideslip such that the 'windward' leading edge is flying at a reduced sweep angle and the 'leeward' leading edge is flying at an increased sweep angle. The lift distribution can be established via separate applications of Diederich's method for each half-airfoil.

3.5.2 Aircraft Rotation

Necessarily an airfoil that is fixed to an aircraft must rotate with the aircraft. The rotation will induce changes in translational velocity of an airfoil, which will vary across the airfoil span. Figure 3.27 shows an airfoil that is offset (aft and up) from a reference centre at which an overall AOA [α] is calculated and at which angular velocities (p, q, r) are defined.

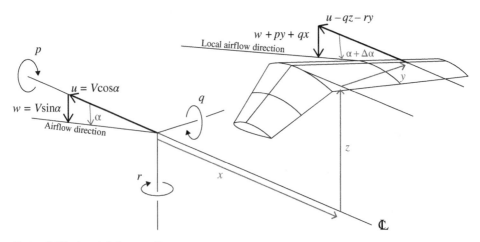

Figure 3.27 Local Reference Frame.

Consider a point on the airfoil quarter-chord with position (x, y, z), using standard conventions for airframe geometry (i.e. x = aft, y = starboard, z = up). The local velocity components become:

$$\begin{pmatrix} u' \\ v' \\ w' \end{pmatrix} = \begin{pmatrix} u \\ v \\ w \end{pmatrix} + \begin{pmatrix} 0 & -r & -q \\ -r & 0 & p \\ q & p & 0 \end{pmatrix} \begin{pmatrix} x \\ y \\ z \end{pmatrix} \tag{3.60}$$

where velocity components are defined in an axis system that is appropriate for flight dynamics (i.e. x = forward, y = starboard, z = down). Local AOA is defined at the local airfoil position, including an induced AOA [$\Delta\alpha$] that is due to rotation:

$$\tan(\alpha + \Delta\alpha) = \frac{w + py + qx}{u - qz - ry} \tag{3.61}$$

A common simplification assumes small AOA (i.e. relatively high airspeed and relatively modest manoeuvres). Thus, forward velocity [u] is approximately equal to airspeed [V]. Therefore, the distribution of induced AOA can be written as follows:

$$\Delta\alpha(\eta) = \frac{x(\eta)}{V}q + \frac{y(\eta)}{V}p \tag{3.62}$$

where $x(\eta)$ and $y(\eta)$ are the coordinates of the quarter-chord point corresponding with the given spanwise ordinate [η]. So, in this case, the local AOA would be specified as follows:

$$\alpha(\eta) = \alpha + \Delta\alpha(\eta) \tag{3.63}$$

3.5.3 Structural Flexibility

The method for determining the lift distribution takes in to account aerodynamic twist, which is the combination of (i) geometric twist and (ii) variation in the zero-lift AOA. On this basis, the method can be extended to accommodate flexure of the airfoil structure. Deflections can be calculated in several ways but they can always be expressed as a distribution of incremental twist [$\Delta\theta(\eta)$] and incremental bending [$\Delta\varphi(\eta)$]. The overall effect is illustrated in Figure 3.28, with the deflected airfoil overlaid on top of the undeflected airfoil. A set of local xyz-frames are drawn along the quarter-chord line and the viewpoint is such that the torsion can be seen clearly from the frame rotation from root to tip.

As shown in Figure 3.29, each xyz-frame provides the resolving the local velocity vector into its components, thereby enabling the calculation of AOA [α] and AOS [β] at each spanwise location:

$$\tan\alpha(\eta) = \frac{w(\eta)}{u(\eta)} \qquad \sin\beta(\eta) = \frac{v(\eta)}{V(\eta)} \tag{3.64}$$

The orientation of each frame is interpreted in Figure 3.30, in terms of two elementary rotations (i.e. rotations about one of the frame axes; x, y, or z). For the purpose of aerodynamic calculation, the first rotation is performed about the local chord line [x-axis] through an angle $\Delta\varphi(\eta)$. This ensures that the zx-plane is locally perpendicular to the airfoil and thus contains the lift and drag vectors. The second rotation is performed about the inclined y-axis

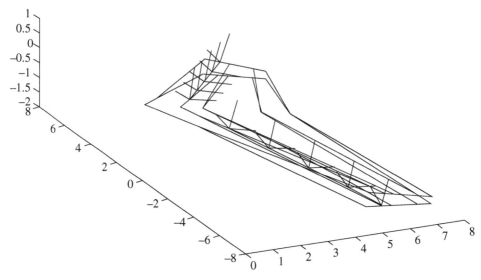

Figure 3.28 Frame Orientation for Flexure.

Figure 3.29 Frame Orientation for Dihedral.

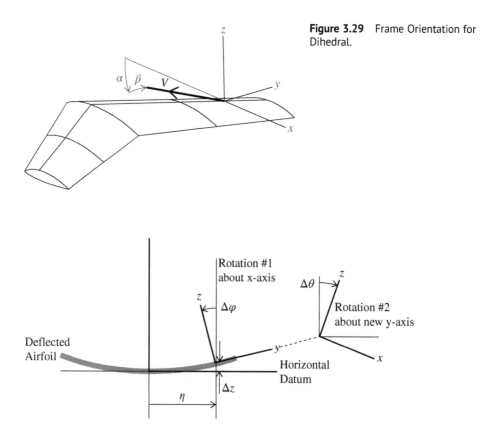

Figure 3.30 Frame Orientation for Flexure.

(that resulted from the first rotation), which is tangential to the airfoil. The angle of rotation $\Delta\theta(\eta)$ represents the torsional deflection. In its final orientation, the local *xyz*-frame provides the basis for calculating local AOA.

3.5.4 Ground Effect

Flight very close to the ground causes a contrition of flow under the wing that is akin to the Bernoulli effect (for flow through a pipe when the pipe diameter is reduced). Flow velocity is increased and downwash is reduced because of the presence of the ground surface. Without going into detail, two factors F_1 and F_2 are applied to the calculations of drag and downwash and to the calculation of lift-curve gradient, respectively:

$$F_1 = 1 + \frac{33\sqrt{(h/b)^3}}{1 + 33\sqrt{(h/b)^3}} \qquad F_2 = 1 + \frac{F_1}{7.469385} \qquad (3.65)$$

where h is the height of the wing above ground and b is the wing span.

3.5.5 Indicial Aerodynamics

Dynamic effects arise from changes in patterns of airflow around the airfoil. These are driven by changes in AOA and in flap deflections. The lift distribution cannot respond instantaneously to such changes. There must a transition between one airflow pattern and another and there must be finite time over which the transition occurs. Also, in general, such changes are continuous, not discrete. This means that the airflow is responding continuously to an AOA profile, which can be thought of a rapid sequence of very small step changes. The overall effect is achieved as an integration of small responses over time. This is *indicial* aerodynamics.

This is a specialised area of research, as it deals with time-varying properties of airflow and its interaction with a moving structure. However, it is important to have some insight into the dynamics of lift growth (as opposed to assuming that the response is so rapid as to be unimportant). This is certainly relevant to the integrated flight/structural dynamics of ultra-large aircraft.

In this context, the Küssner Function $\Psi(\tau)$ and Wagner Function $\Phi(\tau)$ are defined as follows:

$$\Psi(\tau) = 1 - 0.5\varepsilon^{-0.13\tau} - 0.5\varepsilon^{-\tau} \qquad (3.66)$$

$$\Phi(\tau) = 1 - 0.165\varepsilon^{-0.045\tau} - 0.533\varepsilon^{-0.3\tau} \qquad (3.67)$$

where ε is Napier's constant ($\varepsilon \approx 2.718281$) and $\tau = Vt/(c/2)$ is nondimensional time.

These functions account for the lift growth associated with changes in AOA and changes in flap deflection, respectively. The resulting change in the lift distribution will be subject to this growth profile that evolves as a function of the number of half-chord lengths that are traversed at a particular airspeed [V]. It is noted that these equations are strictly applicable to 2D surfaces and that they are only applied to 3D surface on the understanding that they convey a general trend of lift growth.

3.6 Lift Distribution

Diederich's method calculates an approximate lift distribution for general planforms (provided the quarter-chord line is approximately straight). Total lift is calculated by integrating the local lift coefficient $C_l(\eta)$ across the wing span, using a nondimensional length scale $[-1, 1]$ that corresponds with the actual scale $[-b/2, b/2]$:

$$L = \frac{1}{2}\rho V^2 \int_{-1}^{1} c(\eta) C_l(\eta) d\eta \tag{3.68}$$

where the local lift coefficient is defined by:

$$C_l(\eta) = \frac{c'}{c(\eta)} (C_L + \partial_0 C_{l0}(\eta)) \, \partial(\eta) \tag{3.69}$$

where c' = mean geometric chord, $c(\eta)$ = local chord, C_L = 3D lift coefficient, and $C_{l0}(\eta)$ = local lift coefficient at zero-lift.

The distribution is described by the following expressions:

$$\partial_0 = k_0 K_M \cos \Lambda_M \tag{3.70}$$

$$\partial(\eta) = k_1 \left(\frac{c(\eta)}{c'}\right) + k_2 \left(\frac{4}{\pi}\sqrt{1-\eta^2}\right) + k_3 F(\eta) \tag{3.71}$$

where

$$k_1 + k_2 + k_3 = 1$$

The correction function for sweep angle $F(\eta)$ is shown in Figure 3.31 for different values of effective sweep angle Λ_M (as used in Equation 3.16). The coefficients k_0, k_1, k_2, k_3 are

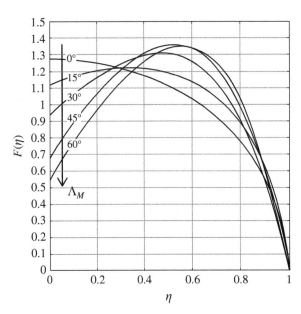

Figure 3.31 Sweep Correction Function.

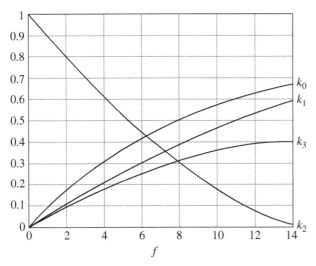

Figure 3.32 Diederich Coefficients.

Table 3.7 Approximations for Diederich Coefficients.

Coefficient	Polynomial Approximation
k_0	$0.26851010(f/14)^3 - 0.90734848f(f/14)^2 + 1.30777777(f/14)$
k_1	$-0.03464646(f/14)^3 - 0.14636363(f/14)^2 + 0.77222222(f/14)$
k_2	$0.18189393(f/14)^3 + 0.27257575(f/14)^2 - 1.44583333(f/14) + 1$
k_3	$-0.14724747(f/14)^3 - 0.12621212(f/14)^2 + 0.67361111(f/14)$

functions of a planform factor $f = (2\pi/a_{2D})(A/14)$, where A is the aspect ratio, a_{2D} is the 2D lift-curve gradient, and Λ_{25} is the quarter-chord sweep angle. These are plotted in Figure 3.32. Numerical approximations are defined in Table 3.7.

Total lift is determined from Equation 3.69:

$$\int c(\eta)C_l(\eta)d\eta = c' \int (C_L + \partial_0 C_{l0}(\eta)) \, \partial(\eta)d\eta$$

$$\int c(\eta)C_l(\eta)d\eta = c'C_L \int \partial(\eta)d\eta + c'\partial_0 \int C_{l0}(\eta)\partial(\eta)d\eta$$

By definition, the total lift is the integral of local lift:

$$\int c(\eta)C_l(\eta)d\eta = c'C_L \tag{3.72}$$

Therefore,

$$\int \partial(\eta)d\eta = 1 \qquad \text{and} \qquad \int C_{l0}(\eta)\partial(\eta)d\eta = 0 \tag{3.73}$$

With reference to Equation 3.80, the first two components of $\partial(\eta)$ integrate to give:

$$\int \left(\frac{c(\eta)}{c'}\right) d\eta = 1 \qquad \int \left(\frac{4}{\pi}\sqrt{1-\eta^2}\right) d\eta = 1 \qquad (3.74)$$

Accordingly, the $F(\eta)$ must be constrained as follows:

$$\int F(\eta) d\eta = 1 \qquad (3.75)$$

For zero total lift, the local lift coefficient is defined by:

$$C_{l0}(\eta) = a\left(\tau(\eta) + \Delta\alpha_0\right) \qquad (3.76)$$

where a is the 3D lift-curve gradient, $\tau(\eta)$ is the aerodynamic twist of the airfoil, and $\Delta\alpha_0$ is the adjustment to the zero-lift AOA in order to counteract the increase in lift due to twist. Diederich formally proposed using the 2D lift-curve gradient in this context but he observed that the 3D lift-curve gradient 'often gives better results'. The aerodynamic twist is defined by:

$$\tau(\eta) = \theta(\eta) - (\alpha_0(\eta) + \alpha_0(0)) \qquad (3.77)$$

where $\theta(\eta)$ = local geometric twist, $\alpha_0(\eta)$ = local zero-lift AOA, and $\alpha_0(0)$ = zero-lift AOA at the wing root.

Equations 3.73 and 3.76 combine as follows:

$$a_{3D}\int \left(\tau(\eta) + \Delta\alpha_0\right)\partial(\eta) d\eta = 0 \qquad (3.78)$$

$$\Delta\alpha_0 \int \partial(\eta) d\eta = -\int \tau(\eta)\partial(\eta) d\eta \qquad (3.79)$$

Recalling Equation 3.74:

$$\Delta\alpha_0 = -\int \tau(\eta)\partial(\eta) d\eta \qquad (3.80)$$

Equation 3.7b gives the total lift coefficient for a wing:

$$C_L = a(\alpha - \alpha_0)$$

where $a = dC_L/d\alpha$ is the 3D lift-curve slope, α is the AOA, and α_0 is the zero-lift AOA. For a generic lift distribution, the zero-lift AOA must take account of the aerodynamic twist distribution. Including this effect, it is necessary to redefine the lift coefficient as follows:

$$C_L = a_{3D}\left(\alpha - \alpha_0 - \Delta\alpha_0\right) \qquad (3.81)$$

Equations 3.69, 3.76, and 3.81 can be combined into a simple expression for the local lift coefficient:

$$C_l(\eta) = a(\eta)\left(\alpha - \alpha_0(\eta)\right) \qquad (3.82)$$

where

$$a(\eta) = \frac{c'}{c(\eta)} a_{3D}\,\partial(\eta) \qquad (3.83)$$

$$\alpha_0(\eta) = \alpha_0 + (1 - \partial_0)\Delta\alpha_0 - \partial_0 \tau(\eta) \qquad (3.84)$$

This applies to a cross-sectional element of the wing but it is stressed that $a(\eta)$ is NOT the 2D lift-curve gradient. It gives the contribution of lift coefficient per unit AOA at a specific wing station to the total lift coefficient per unit AOA.

3.7 Drag Distribution

Lifting Line Theory provides a conceptual framework for understanding finite wing theory and, for current purposes, an efficient scheme for calculating a spanwise drag profile. This is derived by Anderson (2011). Without delving into the detail, this theory expresses lift L in terms of circulation Γ, which is a mathematical concept that implies differential velocity between upper and lower surfaces of the wing.

Accordingly, lift is redefined, as follows:

$$L = \frac{1}{2}\rho V^2 S\, C_L \equiv \rho V \int_{-b/2}^{b/2} \Gamma(y)\, dy \tag{3.85}$$

The downwash at spanwise position y due to the wake is:

$$w(y) = \frac{1}{4\pi} \int_{-b/2}^{b/2} \frac{1}{y - y'} \left(\frac{d\Gamma}{dy}\right)_{y = y'} dy' \tag{3.86}$$

A typical downwash lift distribution is shown in Figure 3.33. Note that the combination of upwash and downwash results in vortex formation around the wing tips. The tip vortices are a significant component of induced drag. The total induced drag is estimated by this method:

$$D_i = -\rho \int_{-b/2}^{b/2} w(y)\Gamma(y)\, dy \tag{3.87}$$

Figure 3.33 Wing Downwash and Tip Vortices.

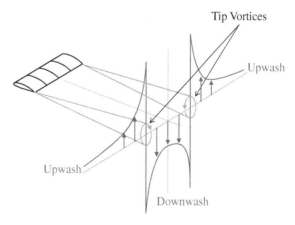

This result will have limited validity because the underlying physical concept is so simple. In particular, wing sweep is known to be problematic. The extension of this method to encompass swept wings involves a more complicated mathematical derivation. This is Extended Lifting Line Theory, also known as Weissinger's theory. Interested readers can refer to Baldoino and Bodstein (2004).

As stated previously, this theory expresses lift L in terms of circulation Γ, which implies differential velocity between upper and lower surfaces of the wing. The downwash associated with a swept wing is expressed as:

$$w(y) = \frac{1}{4\pi} \int_{-b/2}^{b/2} \frac{F(x_P, y, y')\Gamma(y')}{y - y'} \, dy' + \frac{1}{4\pi} \int_{-b/2}^{b/2} F(x_P, y, y')\Gamma(y') \, dy' \tag{3.88}$$

where

$$F(x_P, y, y') = 1 + \frac{x_P(y) - x_C(y')}{\left[(x_P(y) - x_C(y'))^2 + (y - y')^2\right]^{1/2}} \tag{3.89}$$

$$K(x_P, y, y') = \frac{x_P(y) - x_C(y')}{\left[(x_P(y) - x_C(y'))^2 + (y - y')^2\right]^{3/2}} \tag{3.90}$$

Here, the subscripts 'C' and 'P' denote points on the quarter-chord line and three-quarter-chord line, respectively. Without addressing the whys and wherefores, these correspond with the line along which vortices are considered to be bound to the wing and the line along which the downwash is calculated.

4

Longitudinal Flight

4.1 Introduction

4.1.1 Flight with Wings Level

Longitudinal flight refers to straight-line flight with wings level, i.e. the aircraft is not changing course. Thus, the aerodynamics are symmetric and aircraft motion is fully described in the vertical plane. There is a well-established formulary for undertaking calculations that are very familiar to engineers and students alike. Although this does not convey physical understanding, it does indicate clearly how the aircraft geometry affects the generation of the total aerodynamic force/moment system. It is convenient to apply because the effects of cross-flow (or sideslip), rolling motion and yawing motion are not included. In practical terms, this implies that aircraft has 'perfect' stability in its degrees of freedom that are out with the plane of symmetry.

4.1.2 What Chapter 4 Includes

Chapter 4 includes:

- Aerodynamic Fundamentals
- Geometry
- Wing/Body Combination
- All-Moving Tail
- Flight Trim
- Flight Stability
- Trim Drag
- Steady-State Flight Performance
- Dynamic Modes

4.1.3 What Chapter 4 Excludes

Arguably, Chapter 4 does not exclude anything that is essential to explaining whole-aircraft flight. Inevitably, there is more detail available to expand the treatment given here [e.g. Raymer, Pamadi].

4.1.4 Overall Aim

Chapter 4 should provide a comprehensive introduction to aircraft aerodynamics plus flight properties (trim, stability, and performance), together with an overview of longitudinal dynamics. A lot of theory is presented, which enables extensive calculations contributing to preliminary analysis of an aircraft configuration. The calculation of aircraft aerodynamics provides the basis for an aircraft model while the calculation of flight properties and dynamic modes predicts how that model will behave.

4.2 Aerodynamic Fundamentals

Chapter 3 summarises the empirical approach to calculating the aerodynamic characteristics of wings (which deals with the wing and tail surfaces of a conventional aircraft configuration). As a convenient starting point for the chapter, the essential aerodynamic principles are summarised, as follows:

- The quarter-chord point coincides with its aerodynamic centre when in subsonic flight.
- The aerodynamic centre is the centre of induced lift.
- Induced lift is the lift due to angle of attack (AOA) α.
- AOA is the angle between the airflow and the wing chord line.
- The lift coefficient is composed as $C_L = C_{L0} + C_{L\alpha}\alpha$, where C_{L0} is the lift coefficient at zero AOA and $C_{L\alpha}$ is the so-called lift-curve gradient (which is also written as a).
  - Equivalently, $C_L = C_{L\alpha}(\alpha - \alpha_0)$, where α_0 is the AOA that corresponds with zero lift.
- The drag coefficient is composed as $C_D = C_{D0} + C_{Dw} + k_D C_L^2$, where C_{D0} is the profile drag coefficient, C_{Dw} is the wave drag coefficient and k_D is the induced drag factor.
  - Equivalently, $C_D = C_{Dmin} + C_{Dw} + k_D C_{Li}^2$, where C_{Dmin} is the minimum drag coefficient and C_{Li} is the corresponding (ideal) lift coefficient.
- The pitching moment coefficient is composed as $C_M = C_{M0} + C_{M\alpha}\alpha$.
- The pitching moment coefficient calculated at the aerodynamic centre is $C_M = C_{M0}$.

4.3 Geometry

The planform of a typical wing is shown in Figure 4.1. This defines the centre-line chord c_0, the tip chord c_1, and the mean aerodynamic chord $\bar{c}$, as well as sweep angles for the leading edge Λ_0, trailing edge Λ_{100}, and the quarter-chord line Λ_{25}. Geometry parameters are calculated according to the formulary in Table 4.1.

An idealised aircraft is shown in Figure 4.2. In this example, geometry is measured with respect to an **xyz** axis system located at the intersection of the aircraft centre line, the horizontal datum, and the traverse datum. For simplicity, the lifting surfaces have a rectangular planform with span b, chord length c, area $S = bc$, and aspect ratio $A = b/c$. Subscripts W and T denote wing and tail, respectively. Reference points are defined at the quarter-chord positions on each surface, at which aerodynamic forces and moments are calculated. The distance between these points is l. The product of area and distance is the *tail volume* ($V_T = lS_T$), which determines the effectiveness of the tail as a producer of pitching moment.

Figure 4.1 Typical Wing.

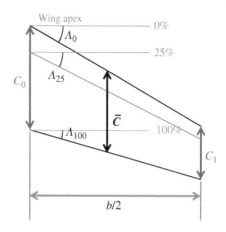

Table 4.1 Wing Geometric Parameters.

Taper ratio	$\lambda = c_1/c_0$	Mean geometric chord	$c' = \dfrac{1+\lambda}{2}c_0$
Aspect ratio	$A = b/c'$	Mean aerodynamic chord	$\bar{c} = \left(\dfrac{2}{3}\right)\dfrac{1+\lambda+\lambda^2}{1+\lambda}c_0$
Planform area	$S = bc'$	Wetted area	$S_{wet} = 2S\left(1 + \dfrac{1}{4}\tau_0\dfrac{1+\lambda\tau_1/\tau_0}{1+\lambda}\right)$
Thickness ratio	$\tau = \dfrac{t}{c}$	Sweep angle of $n\%$ chord line	$\tan\Lambda_n = \tan\Lambda_0 - \dfrac{n}{100}\left(\dfrac{4}{A}\right)\dfrac{1-\lambda}{1+\lambda}$

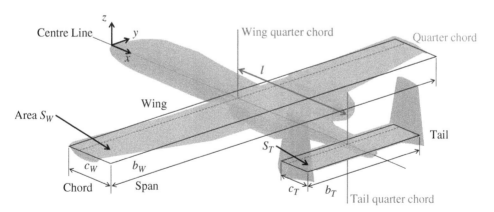

Figure 4.2 Idealised Aircraft.

Figure 4.3 then shows a more typical geometry associated with commercial transport aircraft, indicating the exposed or wetted area of the wing S'_W (as opposed to the total area S_W) and the fuselage fineness ratio (which is the ratio of length l_B to width w_B). It also indicates the maximum cross-sectional area of the nose S_N.

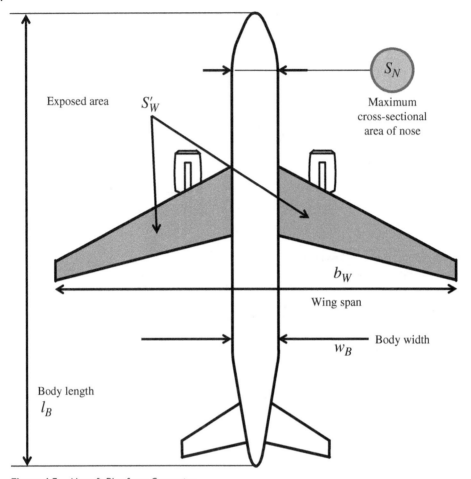

Figure 4.3 Aircraft Planform Geometry.

4.4 Wing/Body Combination

4.4.1 Lift Force

The wing lift-curve gradient was defined by Equation 3.16. In this section, it is written as:

$$a_W = \frac{2\pi A}{2 + \sqrt{4 + (A/\kappa)^2 K_M^2 (1 + \tan \Lambda_M)^2}} \tag{4.1}$$

where

$$K_M = \sqrt{1 - M_N^2} \qquad \text{and} \qquad \tan \Lambda_M = \frac{\tan \Lambda_{50}}{K_M}$$

and where M_N is the flight Mach number, A is the aspect ratio, and Λ_{50} is the mid-chord sweep angle. The parameter κ is the ratio of the 2D lift-curve gradient to the ideal lift-curve gradient, as defined by Equation 3.21. The most commonly quoted extension for wing/body aerodynamics is:

$$a_{WB} = 2K_S \frac{S_N}{S_W} + a_W K_{WB} \frac{S'_W}{S_W} \tag{4.2}$$

where

$$K_S = 1 - \exp\left(-\frac{1}{4}\left(\frac{l_B}{w_B} + 2.1\right)\right) \qquad K_{WB} = \left(1 + \frac{w_B}{b_W}\right)^2 \qquad S_N = \pi\left(\frac{w_B}{2}\right)^2$$

and where the geometric parameters are defined in Figure 4.3.

The calculation of lift coefficient is undertaken, as follows:

$$C_L = a_{WB}(\alpha - \alpha_0) + \Delta C_L \tag{4.3}$$

where ΔC_L is the incremental change in wing/body lift due to the vertical position of the wing:

$$\Delta C_L = K_z\left(\frac{w_B}{b_W}\right) \tag{4.4}$$

where K_z is 0.1 for a high wing, 0 for a mid wing and −0.1 for a low wing.

4.4.2 Downwash

Airflow around the wing is deflected upwards ahead of the leading edge (known as 'upwash') and downwards behind the trailing edge (known as 'downwash') as shown in Figure 4.4. The variation of downwash with AOA can be written as:

$$\varepsilon = \varepsilon_\alpha(\alpha - \alpha_0) + \Delta\varepsilon \tag{4.5}$$

where α is the AOA and α_0 is the zero-lift AOA, ε_α is the downwash gradient and $\Delta\varepsilon$ is the incremental change in downwash due to flap deflection.

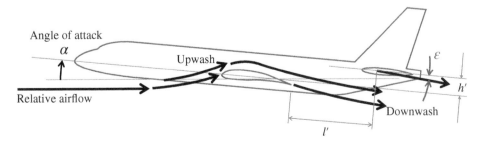

Figure 4.4 Effect of Downwash on Horizontal Tail.

The gradient is given by DATCOM:

$$\varepsilon_\alpha = 4.44 \left(K_A K_T K_\lambda\right)^{1.19} \tag{4.6}$$

where

$$K_A = \frac{1}{A} - \frac{1}{1 + A^{1.7}} \qquad K_T = \frac{1 - |h'/b|}{\sqrt[3]{2l'/b}} \qquad K_\lambda = \frac{10 - 3\lambda}{7} \tag{4.7}$$

The aspect ratio is A and the taper ratio is λ [cf. Table 4.1]. The position at which downwash is being calculated is (l', h'), where l' is measured along the wing chord plane from the trailing edge and h' is measured perpendicular to the wing chord plane.

The effect of flap deflection δ_F can be calculated as:

$$\Delta \varepsilon = \varepsilon_\delta \delta_F \tag{4.8}$$

In its simplest (and most commonly quoted) form:

$$\varepsilon_\delta = \varepsilon_\alpha \alpha_\delta \tag{4.9}$$

where ε_α is the downwash gradient and α_δ is the flap effectiveness [cf. Figure 4.5]:

$$\alpha_\delta = 1 - \frac{(\varphi - \sin \varphi)}{\pi} \qquad \text{where} \qquad \cos \varphi = 2 \left(\frac{c_F}{c}\right) - 1 \tag{4.10}$$

where c is the wing chord length and c_F is the flap chord length. A slightly less simple calculation is:

$$\varepsilon_\delta \delta = \frac{K_\varepsilon}{A} \left(\frac{b_W}{w_B}\right) \Delta C_L \tag{4.11}$$

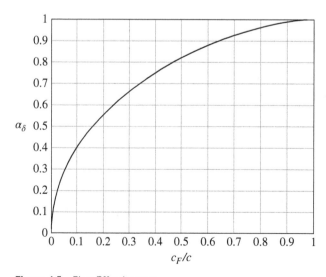

Figure 4.5 Flap Effectiveness.

where A is the wing aspect ratio, b_W is the wing span, w_B is the body width and ΔC_L is the change in lift due to flap deflection. The calibration factor is given by:

$$K_\varepsilon = 49.5726\sigma^2 - 4.0427\sigma + 21 \tag{4.12}$$

where $\sigma = h'/b_W$. Recalling Section 3.4.1, the incremental lift is given by:

$$\Delta C_L = \left(K_b(\eta_{ob}) - K_b(\eta_{ib})\right) K_c K_F \frac{a}{a_{2D}} C_{L\delta,theory} K_\delta \delta_F \tag{4.13}$$

where all parameters were defined for Equations 3.44 and 3.45.

In addition, there is a time delay $t' = l_T/V$ for the air to flow downstream from the wing, such that:

$$\varepsilon(t) = \varepsilon_0 + \varepsilon_\alpha \alpha(t - t') + \varepsilon_\delta \delta_F(t - t') \tag{4.14}$$

where $\varepsilon_0 = -\varepsilon_\alpha \alpha_0$.

Using power series expansions, this becomes:

$$\varepsilon(t) \approx \varepsilon_0 + \varepsilon_\alpha(\alpha(t) - t'\dot{\alpha}(t)) + \varepsilon_\delta\left(\delta_F(t) - t'\dot{\delta}_F(t)\right) \tag{4.15}$$

Often, the rate of change of flap deflection is ignored because it is relatively slow, in which case:

$$\varepsilon(t) \approx \varepsilon_0 + \varepsilon_\alpha(\alpha(t) - t'\dot{\alpha}(t)) + \varepsilon_\delta \delta_F(t) \tag{4.16}$$

4.4.3 Pitching Moment

The contribution of the fuselage body to pitching moment is most easily calculated using the Munk–Multhopp approximation that was originally developed for airships and subsequently refined for aircraft. The method divides the body into segments forward and aft of the wing root chord and then aggregates the aerodynamic influences of upwash and downwash, respectively, as shown in Figure 4.4. An example of fuselage partitioning is shown in Figure 4.6, numbered 1 to 20, with a gap over the wing root chord because the flowfield in this region has very little effect on pitching moment.

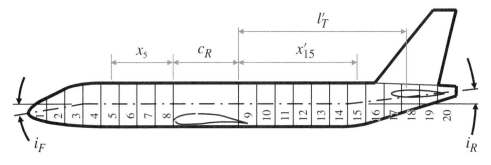

Figure 4.6 Fuselage Partition for Multhopp's Method.

The evaluation of C_{M0} and $C_{M\alpha}$ for the fuselage body is performed as follows:

$$C_{M0} = \frac{\pi}{2} \frac{K_2 - K_1}{S_W \bar{c}_W} \sum_n w_n^2 (\alpha_{W0} + i_n) \Delta x_n \qquad (4.17)$$

$$C_{M\alpha} = \frac{\pi}{2} \frac{1}{S_W \bar{c}_W} \sum_n w_n^2 \frac{\partial \varepsilon_U}{\partial \alpha} \Delta x_n \qquad (4.18)$$

where w_n is the width of the nth fuselage segment, i_n is the incidence of the fuselage camber line at the centre point of the segment, and Δx_n is the longitudinal thickness of the segment. The symbols S_W and $\bar{c}_W$ denote the total area and the mean aerodynamic chord of the wing, as defined already. In the example, the forward fuselage camber is inclined at angle i_F (for Segments 1–3) and the rear fuselage camber is inclined at angle i_R (for Segments 15–20). The camber line is parallel with the fuselage datum for Segments 4–14.

The combination $K_2 - K_1$ is the correction factor for the fuselage fineness ratio, which is shown in Figure 4.7. The graph of $\partial \varepsilon_U / \partial \alpha$ is shown in Figure 4.8 as two functions of x/c_R, where x is the longitudinal separation between the mid-point of each segment and the leading edge of the wing root chord and c_R is the length of the root chord. In the example, $\partial \varepsilon_U / \partial \alpha$ for Segments 1–7 is obtained from Curve A and, for Segment 8, it is obtained from Curve B. The downwash gradient, $\partial \varepsilon / \partial \alpha$, behind the trailing edge of the root chord is obtained from:

$$\frac{\partial \varepsilon_U}{\partial \alpha} = \frac{x'}{l'_T} \left(1 - \frac{\partial \varepsilon}{\partial \alpha} \right) \qquad (4.19)$$

where x' is the separation between the mid-point of each segment and the trailing edge of the wing root chord and l'_T is the corresponding distance to the tail aerodynamic centre.

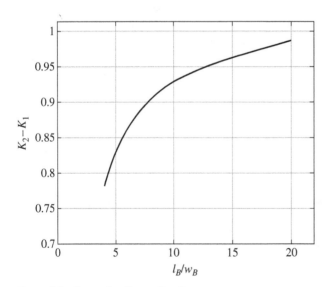

Figure 4.7 Correction Factor $K_2 - K_1$.

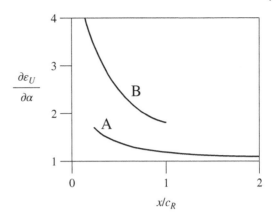

Figure 4.8   Upwash Gradient.

4.4.4   Aerodynamic Centre

The wing/body lift-curve gradient is $C_{La} = a_{WB}$ (from Equation 4.2) and the pitching moment gradient is C_{Ma} (from Equation 4.16). Thus, The aerodynamic centre is located at distance x_{WB} from the aircraft datum point:

$$\frac{X_{WB}}{\bar{c}_W} = \frac{C_{Ma}}{C_{La}} \tag{4.20}$$

where $\bar{c}_W$ is the mean aerodynamic chord of the wing.

4.4.5   Drag Force

Following the same principles as Chapter 3, the aircraft drag coefficient is calculated as:

$$C_D = C_{D0} + C_{Di} + C_{Dw} \tag{4.21}$$

where $\rho V^2/2$ is the dynamic pressure, S is the panform area, and where C_{D0}, C_{Di}, and C_{Dw} are the coefficients of profile drag, induced drag, and wave drag, respectively.

Fuselage wave drag is difficult to address using semi-empirical methods because so little has been published on this subject. So, the standard approach is to assume that the dominant effect is due to shock formation over the wing (and refer to Chapter 3).

Induced drag is calculated as for an isolated wing but based on wing/fuselage lift coefficient:

$$C_{Di} = \frac{C_L^2}{\pi Ae} \tag{4.22}$$

where $C_L = a_{WB}(\alpha - \alpha_0)$.

Profile drag was discussed in Chapter 3 for an isolated wing and the same method is applied here for the whole aircraft (as a summation of drag for all components), such that:

$$C_{D0} = \sum_{components} Q\, C_f\, F \frac{S_{wet}}{S'_W} \tag{4.23}$$

Generically, Q is the interference factor between components, C_f is the flat-plate skin friction coefficient, F is the form factor, S_{wet} is the wetted surface area, and S'_W to the exposed planform area of the wing.

The interference factors are estimated as $Q = 1.00$ for wing and body, $Q = 1.03$ for horizontal and vertical tail surfaces, $Q = 1.30$ for engines and $Q = 1.50$ for engine pylons.

The flat-plate skin friction coefficient is:

$$C_f = \lambda\, C_{f,laminar} + (1 - \lambda)\, C_{f,turbulent} \tag{4.24}$$

where λ is fraction of the surface length that has laminar flow [where it is assumed that $\lambda \approx 0.3$ for the wing, $\lambda \approx 0.2$ for the horizontal and vertical tail surfaces, $\lambda \approx 0.1$ for a fuselage, and $\lambda \approx 0$ for an engine nacelle]. The laminar and turbulent flow coefficients are:

$$C_{f,laminar} = \frac{1.328}{\sqrt{Re}} \qquad C_{f,turbulent} = \frac{0.455}{(log\, Re)^{2.58}\left(1 + 0.144 M_N^2\right)^{0.65}} \tag{4.25}$$

where M_N is Mach number and Re is Reynolds Number.

Recall from Chapter 3 that the value of Reynolds Number is limited for subsonic aerodynamics:

$$Re \leq 38.21 \left(\frac{c}{k}\right)^{1.053} \tag{4.26}$$

where c is the chord length and k is the skin roughness factor (0.634×10^{-5} m for smooth paint).

The form factor for a wing is given by:

$$F = 1.34 \left[1 + 0.6\left(\frac{c}{x}\right)\tau + 100\tau^2\right] M^{0.18} (\cos \Lambda_m)^{0.28} \tag{4.27}$$

where x/c gives the location of the maximum thickness ratio τ as a fraction of the wing chord (typically 0.3–0.5). The angle Λ_m is the sweep angle at the line of maximum thickness. The wetted area is estimated in Table 4.1:

$$S_{wet} = 2S\left(1 + \frac{1}{4}\tau_0 \frac{1 + \lambda \tau_1/\tau_0}{1 + \lambda}\right) \tag{4.28}$$

The form factor of the fuselage is:

$$F = 1 + \frac{60}{f^3} + \frac{f}{400} \tag{4.29}$$

where $f = l_F/d_F$, in which l_F is fuselage length and d_F is fuselage diameter. The wetted area is:

$$S_{wet_F} = \pi \lambda_F \left(1 - \frac{2}{\lambda_F}\right)^{2/3}\left(1 + \frac{1}{\lambda_F^2}\right) \tag{4.30}$$

where $\lambda_F = l_F d_F$.

The geometric parameters for a typical engine nacelle are given in Figure 4.9. The form factor is:

$$F = 1 + \frac{0.35}{f} \tag{4.31}$$

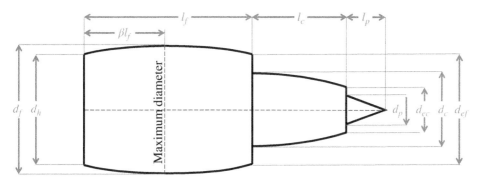

Figure 4.9 Engine Nacelle Geometry.

where $f = l_E/d_E$ in which $l_E = l_f + l_c + l_p$ and $d_E = d_f$. The wetted area is estimated as:

$$S_{wet_E} = S_{wet_{Ef}} + S_{wet_{Ec}} + S_{wet_{Ep}} \tag{4.32}$$

where subscripts are 'E' for the engine, 'Ef' for the fan nacelle, 'Ec' for the core nacelle, and 'Ep' for the plug. The calculations are:

$$S_{wet_{Ef}} = \pi l_f d_f \left[2 + 0.35\beta + 0.8\beta \frac{d_h}{d_f} + 1.15(1-\beta)\frac{d_{ef}}{d_f} \right] \tag{4.33}$$

$$S_{wet_{Ec}} = \pi\, l_c d_c \left[1 - \frac{1}{3}\left(1 - \frac{d_{ec}}{d_c}\right)\left(1 - 0.18\left(\frac{d_c}{l_c}\right)^{5/3}\right) \right] \tag{4.34}$$

$$S_{wet_{Ep}} = 0.7\,\pi\, l_p d_p \tag{4.35}$$

4.5 All-Moving Tail

4.5.1 Lift Force

The calculation of lift coefficient is undertaken for the horizontal tail, as follows:

$$C_L = a_T \left(\alpha + \frac{l}{V}q + \delta_T - \varepsilon \right) \tag{4.36}$$

where a_T is the lift-curve gradient for the tail, α is the aircraft AOA, q is the aircraft pitch rate, δ_T is the tail rotation, and ε is the downwash at the tail (as given by Equation 4.16). Also, l is defined in Figure 4.3 (as the longitudinal separation of wing and tail) and V is the airspeed. Thus, this calculation becomes:

$$C_L = a_T \left((1-\varepsilon_\alpha)\alpha + \frac{l}{V}q + \delta_T - \varepsilon_0 + \varepsilon_\alpha t'\dot{\alpha} + \varepsilon_\delta \delta_F \right)$$

This is de-cluttered as follows:

$$C_L = a_T'\alpha + a_T q' + a_T \delta_T + a_T \alpha' \tag{4.37}$$

where the newly introduced symbols are:

$$a_T' = a_T(1 - \varepsilon_\alpha) \qquad q' = q\left(\frac{l}{V}\right) \qquad \alpha' = -\varepsilon_0 + \varepsilon_\alpha t'\dot\alpha + \varepsilon_\delta \delta_F$$

The tail lift-curve gradient was defined by Equation 3.16. In this section, it is written as:

$$a_T = \frac{2\pi A}{2 + \sqrt{4 + (A/\kappa)^2 K_M^2 (1 + \tan \Lambda_M)^2}} \qquad (4.38)$$

where

$$K_M = \sqrt{1 - M_N^2} \qquad \text{and} \qquad \tan \Lambda_M = \frac{\tan \Lambda_{50}}{K_M}$$

and where M_N is the flight Mach number, A is the aspect ratio, and Λ_{50} is the mid-chord sweep angle. The parameter κ is the ratio of the 2D lift-curve gradient to the ideal lift-curve gradient, as defined by Equation 3.21.

4.5.2 Pitching Moment

Typically, tail surfaces are uncambered. Therefore, the pitching moment about the aerodynamic centre (which is constant) is identically zero.

4.5.3 Drag Force

Following Section 4.4.3, the tail drag coefficient is calculated as:

$$C_D = C_{D0} + C_{Di} + C_{Dw} \qquad (4.39)$$

where $\rho V^2/2$ is the dynamic pressure, S is the panform area, and where C_{D0}, C_{Di}, and C_{Dw} are the coefficients of profile drag, induced drag, and wave drag, respectively. These contributions are calculated according to the methods set out already.

4.6 Flight Trim

Figure 4.10 shows an aircraft in straight-and-level flight for a given velocity V and AOA α. Lift vectors are applied on the wing and tail. The aircraft Centre of Gravity (CG) is shown as a circle with alternate black and white quadrants. Engine thrust will contribute to the vertical force acting on the aircraft but, for small AOA, it is convenient to neglect it. It will also contribute a pitching moment because the thrust line typically lies above or (usually) below the wing chord datum.

This aerodynamic force/moment system is evaluated with respect to the CG, as follows:

$$L = L_{WB} + L_T \qquad M = M_0 + l_{WB}L_{WB} - l_T L_T \qquad (4.40)$$

where subscripts WB and T denote wing/bodyand tail, respectively. The moment M_0 is the pitching moment of the wing/body calculated at its aerodynamic centre. The distances l_{WB} and l_T define the aerodynamic centres of the wing/body and tail relative to the CG.

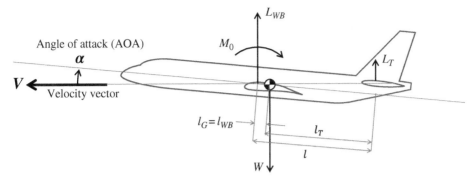

Figure 4.10 Aircraft in Straight-and-Level Flight.

The aircraft is trimmed (i.e. is in a balanced flight condition) when $L = W$ and $M = 0$. After some manipulation, this is determined as:

$$L_{WB} = \frac{l_T}{l}W - \frac{1}{l}M_0 \qquad L_T = \frac{l_{WB}}{l}W + \frac{1}{l}M_0 \qquad (4.41)$$

4.7 Flight Stability

Figure 4.11 shows the incremental changes in wing lift and tail lift when AOA is changed by a small angle $\Delta\alpha$, while the tail deflection δ_T is fixed. The combined effect can be calculated at a point N, as defined in the figure. From Equation 4.40, the aerodynamic force/moment system is:

$$L = L_{WB} + L_T \qquad\qquad M = M_0 + l_{WB}L_{WB} - l_T L_T$$

An alternative formulation is given by:

$$M = M_0 + l_G(L - L_T) - (l - l_G)L_T$$
$$M = M_0 + l_G(L - L_T) - lL_T$$

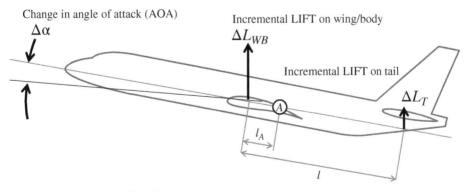

Figure 4.11 Incremental Lift Force and Pitching Moment.

where l_{WB} is renamed l_G [cf. Figure 4.10], as the distance to the CG from the aircraft reference point. This is rationalised to give:

$$M = M_0 + l_{WB}L - lL_T \tag{4.42}$$

The total change in lift force and pitching moment is given by:

$$\Delta M = l_G \Delta L - l \Delta L_T \tag{4.43}$$

The criterion for pitch stability is that $\Delta M < 0$ when $\Delta M > 0$ and vice versa, i.e. an increase or decrease in AOA produces a restoring moment. The transition between stability and instability occurs when $\Delta M = 0$.

Recalling Equation 3.12, the location of the aerodynamic centre is $x_A \approx -\Delta M/\Delta L$, measured from the CG (because the lift force and the pitching moment in Equation 4.43 are calculated at the CG):

$$x_A \approx l_A - l_G \tag{4.44}$$

where

$$l_A = l \frac{\Delta L_T}{\Delta L} \tag{4.45}$$

Clearly, if $l_A = l_G$, then $\Delta M = 0$ and the aircraft has neutral stability. In this context, the aerodynamic centre of the complete aircraft is shown as point 'A' in Figure 4.11.

Following on from earlier discussions, the lift forces generated by the complete aircraft and the tail are calculated, respectively, as follows:

$$L = QS_W a(\alpha - \alpha_0) \qquad L_T = QS_T(a_T'\alpha + a_T q' + a_T \delta_T + a_T \alpha') \tag{4.46}$$

where $Q = \rho V^2/2$ is the dynamic pressure, S_W is the wing reference area, S_T is the tail reference area, and α is the aircraft AOA. The lift-curve gradients are written as a for the aircraft and a_T for the tail. The remaining terms relate to downwash [cf. Equation 4.37]:

$$a_T' = a_T(1 - \varepsilon_\alpha) \qquad q' = q\left(\frac{l}{V}\right) \qquad \alpha' = -\varepsilon_0 + \varepsilon_\alpha t'\dot{\alpha} + \varepsilon_\delta \delta_F$$

By definition, for a steady-state flight condition with fixed controls (δ_T and δ_F) and constant AOA ($\dot{\alpha} = 0$), the incremental lift forces generated by the aircraft are:

$$\Delta L = QS_W a \Delta\alpha \qquad \Delta L_T = QS_T\left(a_T'\Delta\alpha + \frac{l}{V}a_T\Delta q\right) \tag{4.47}$$

The lift force required to maintain a pitch rate q can be written as:

$$mVq = QS_W a(\alpha - \alpha_0) \tag{4.48}$$

An incremental change in pitch rate is equivalent to a change in AOA, as follows:

$$\Delta q = \frac{QS_W a}{mV}\Delta\alpha = \frac{Va}{\mu\bar{c}}\Delta\alpha \tag{4.49}$$

where nondimensional mass μ is defined as:

$$\mu = \frac{m}{(\rho/2)S\bar{c}} \tag{4.50}$$

Thus, the incremental lift forces become:

$$\Delta L = QS_W a\,\Delta\alpha \qquad \Delta L_T = QS_T\left(a'_T + \frac{l}{V}a_T\frac{Va}{\mu\bar{c}}\right)\Delta\alpha \tag{4.51}$$

Now, the position of the aerodynamic centre (relative to the aircraft datum point) is given by Equation 4.45. Applying substitutions from Equation 4.51, this becomes:

$$l_A = l\frac{\Delta L_T}{\Delta L} = l\frac{S_T}{S_W}\frac{a'_T}{a} + l\frac{S_T}{S_W}\frac{la_T}{\mu\bar{c}_W} \tag{4.52}$$

It is convenient to use a length scale that is nondimensionalised with respect to the wing mean aerodynamic chord $\bar{c}_W$. This gives the aerodynamic centre as:

$$h_A = \frac{l}{\bar{c}_W}\frac{\Delta L_T}{\Delta L} = \overline{V}_T\frac{a'_T}{a} + \overline{V}_T\frac{la_T}{\mu\bar{c}} \tag{4.53}$$

where the relative tail volume is defined by:

$$\overline{V}_T = \frac{lS_T}{\bar{c}_W S_W} \tag{4.54}$$

Re-stating the CG position as $h_G = l_G/\bar{c}$, the associated stability margin is defined as:

$$H_A = h_A - h_G \tag{4.55}$$

There are two conditions that are considered when establishing a stability margin of this type. Firstly, when pitch rate is zero ($q = 0$), the aerodynamic centre is called the neutral point and is defined by:

$$h_N = \overline{V}_T\frac{a'_T}{a} \tag{4.56}$$

The associated stability margin is called the 'controls-fixed static margin' and is defined by:

$$H_N = h_N - h_G \tag{4.57}$$

Secondly, when pitch rate is non-zero ($q \neq 0$), the aerodynamic centre is called the manoeuvre point and is defined by:

$$h_M = h_N + \overline{V}_T\frac{la_T}{\mu\bar{c}} \tag{4.58}$$

The associated stability margin is called the 'controls-fixed manoeuvre margin' and is defined by:

$$H_M = h_M - h_G \tag{4.59}$$

Considering straight-and-level flight with fixed controls, the incremental lift forces generated by the wing/body combination and the tail are calculated, respectively, as follows:

$$\Delta L_{WB} = QS_W a_{WB}\, \Delta\alpha \qquad\qquad \Delta L_T = QS_T a'_T\, \Delta\alpha \qquad\qquad (4.60)$$

where $Q = \rho V^2/2$ is the dynamic pressure, S_W is the wing reference area, S_T is the tail reference area, and α is the aircraft AOA. The lift-curve gradients are written as a_{WB} for the wing/body and a_T for the tail. In this case, the neutral point is established, as follows:

$$l_N = l\frac{\Delta L_T}{\Delta L_{WB} + \Delta L_T} = l\,\frac{S_T a'_T}{S_W a_{WB} + S_T a'_T} \qquad\qquad (4.61)$$

Assuming that the lift-curve gradients are similar, a very rough approximation is:

$$l_N \approx l\frac{S_T}{S_W + S_T} \qquad\qquad (4.62)$$

4.8 Trim Drag

4.8.1 Minimum Drag

For equilibrium in level flight, $L = W$, where L is lift and W is weight. The lift coefficient is given by:

$$C_L = \frac{W}{QS_W} \qquad\qquad (4.63)$$

where $Q = \rho V^2/2$ is the dynamic pressure and S_W is the wing area. The corresponding drag coefficient at low airspeed is given by:

$$C_D = C_{D0} + K_D C_L^2 \qquad\qquad (4.64)$$

where C_{D0} is the zero-lift drag coefficient and K_D is the induced drag factor for the aircraft. Combining these equations:

$$D = QS\left(C_{D0} + K_D\left(\frac{W}{QS_W}\right)^2\right)$$

$$D = QSC_{D0} + \frac{K_D}{QS_W}W^2 \qquad\qquad (4.65)$$

Thus, in this simple theory, the drag calculation for a given weight and a given density (which is determined by altitude) has two components, one that varies with V^2 and one that varies with V^{-2}. This is shown in Figure 4.12. Clearly, the addition of these components gives a drag curve that reveals a minimum drag speed. Differentiating Equation 4.65:

$$\frac{dD}{dQ} = SC_{D0} - \frac{K_D W^2}{Q^2 S}$$

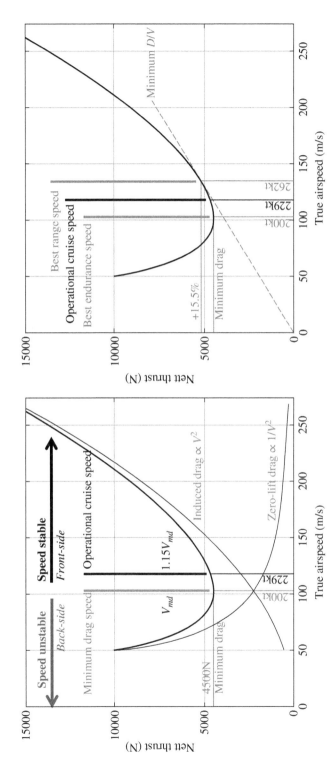

Figure 4.12 Typical Trim Drag Curve (left panel), Typical Trim Drag Curve (Right Panel).

Minimum drag D_{min} and the corresponding dynamic pressure Q_{md} occur when $dD/dQ = 0$:

$$Q_{md} = \sqrt{\frac{K_D}{C_{D0}}\left(\frac{W}{S_W}\right)} \qquad D_{min} = 2\sqrt{K_D C_{D0}}\, W \qquad (4.66)$$

The corresponding value of true airspeed V_{md} for a known density ρ is given by:

$$V_{md}^2 = \frac{2}{\rho}\sqrt{\frac{K_D}{C_{D0}}\left(\frac{W}{S_W}\right)} \qquad (4.67)$$

Figure 4.12 shows a minimum drag speed V_{md} equal to 200 kt for the particular aircraft being analysed. For trim conditions above that speed, the gradient of the drag curve is positive and any unwanted speed increase will cause more that will decelerate the aircraft; any unwanted speed decrease will cause less drag that will accelerate the aircraft. So, the speed is stable above V_{md}.

However, below V_{md}, the opposite process applies. Any unwanted speed increase will cause less drag that will accelerate the aircraft and any unwanted speed decrease will cause more drag that will decelerate the aircraft. The speed is unstable below V_{md}.

This distinction is identified by the terms 'front side' (above V_{md}) and 'back side' (below V_{md}). Note that an operational cruise speed is specified at 15% above V_{md}. This is set in order to ensure that the aircraft can be easily held in front-side operation.

The maximum range speed V_{mr} and maximum endurance speed V_{me} are shown in order to put the cruise speed into context. For a jet-powered aircraft, $V_{me} = V_{md}$ and $V_{mr} = \sqrt{3}\,V_{md}$. Note that 'best range' is determined by minimum D/V and implies a 15.5% increase above minimum drag. The derivation is not offered here but is available in many texts, such as Eshelby (2000).

Figure 4.13 also summarises three practical limitations that are associated with trim drag. The *first limitation* is the reduction in the best range speed caused by wave drag. The true airspeed is shown that corresponds with the Critical Mach Number at which the flow first becomes supersonic somewhere over the aircraft. The *second limitation* is the thrust required to maintain a given trim condition. In the example shown, the speed range is between 185 and 478 kt, defined by maximum engine speed. The minimum engine speed for trim is 86%, for which the trim speed is 300 kt. Note that the speed range contracts as altitude increases. The *third limitation* is AOA, which increases as speed decreases and ultimately the aircraft will stall. In the example, this occurs at approximately 215 kt. At low altitude, this problem is resolved by deploying flaps!

4.8.2 Relative Speed and Relative Drag

It is useful to relate airspeed V to the minimum drag speed V_{md} for a given flight condition:

$$u = \frac{V}{V_{md}} \qquad (4.68)$$

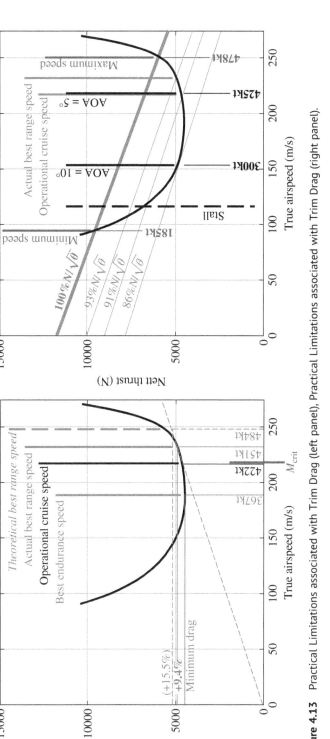

Figure 4.13 Practical Limitations associated with Trim Drag (left panel), Practical Limitations associated with Trim Drag (right panel).

The corresponding ratio of dynamic pressure is obtained as follows:

$$Q = \frac{1}{2}\rho V^2 \qquad Q_{md} = \frac{1}{2}\rho V_{md}^2 \qquad \Longrightarrow \qquad \frac{Q}{Q_{md}} = \left(\frac{V}{V_{md}}\right)^2 = u^2 \tag{4.69}$$

Similarly, it is useful to relate drag D to minimum drag D_{min} for a given flight condition:

$$d = \frac{D}{D_{min}} \tag{4.70}$$

Aircraft drag is defined by Equation 4.65:

$$D = QSC_{D0} + \frac{K_D}{QS}W^2$$

Minimum drag is defined by Equation 4.66:

$$D_{min} = 2\sqrt{K_D C_{D0}}\,W$$

The ratio of these quantities is derived as follows:

$$\frac{D}{D_{min}} = \frac{1}{2}\left(\frac{qS}{W}\frac{C_{D0}}{\sqrt{K_D C_{D0}}} + \frac{W}{qS}\frac{K_D}{\sqrt{K_D C_{D0}}}\right)$$

$$\frac{D}{D_{min}} = \frac{1}{2}\left(\frac{qS}{W}\sqrt{\frac{C_{D0}}{K_D}} + \frac{W}{qS}\sqrt{\frac{K_D}{C_{D0}}}\right) = \frac{1}{2}\left(\frac{Q}{Q_{md}} + \frac{Q_{md}}{Q}\right)$$

The final result is:

$$\frac{D}{D_{min}} = \frac{1}{2}\left(\left(\frac{V}{V_{md}}\right)^2 + \left(\frac{V_{md}}{V}\right)^2\right) \tag{4.71}$$

Using the definitions of relative speed and relative drag in Equations 4.68 and 4.70:

$$d_{min} = \frac{1}{2}\left(u^2 + \frac{1}{u^2}\right) \tag{4.72}$$

4.8.3 Variation of Minimum Drag Speed

From Equations 4.67, the minimum drag speed V_{md} for known density ρ is given by:

$$V_{md}^2 = \frac{2}{\rho}\sqrt{\frac{K_D}{C_{D0}}}\left(\frac{W}{S}\right) = \frac{2}{S}\sqrt{\frac{K_D}{C_{D0}}}\left(\frac{W}{\rho}\right)$$

Definitions of relative pressure δ, relative temperature θ, and relative density σ were given in Equation 1.16:

$$\delta = \frac{P}{P_0} \qquad \theta = \frac{T}{T_0} \qquad \sigma = \frac{\rho}{\rho_0}$$

where the subscript 0 denotes sea level. The Ideal Gas Law can be written (as in Equation 1.17):

$$\delta = \sigma\theta$$

Using these definitions, the minimum drag speed can be re-expressed as:

$$V_{md}^2 = \frac{2}{\rho_0 S}\sqrt{\frac{K_D}{C_{D0}}}\left(\frac{W}{\sigma}\right) \qquad \left(\frac{V_{md}}{\sqrt{\theta}}\right)^2 = \frac{2}{\rho_0 S}\sqrt{\frac{K_D}{C_{D0}}}\left(\frac{W}{\delta}\right) \tag{4.73}$$

The minimum drag Mach number is:

$$M_{md} = \frac{V_{md}}{v} = \frac{V_{md}}{v_0\sqrt{\theta}} \tag{4.74}$$

where v denotes the speed of sound. Combining these equations:

$$M_{md}^2 = \frac{2}{a_0^2\rho_0 S}\sqrt{\frac{K_D}{C_{D0}}}\left(\frac{W}{\delta}\right) \tag{4.75}$$

Therefore, the variation of minimum drag speed can be summarised as:

$$Q_{md} \propto W \qquad V_{md} \propto \sqrt{\frac{W}{\sigma}} \qquad \frac{V_{md}}{\sqrt{\theta}} \propto \sqrt{\frac{W}{\delta}} \qquad M_{md} \propto \sqrt{\frac{W}{\delta}} \tag{4.76}$$

4.8.4 Minimising 'Minimum' Drag

For straight-and-level flight, Equation 4.41 gives the lift forces for wing/body and tail. Hypothetically, if the pitching moment M_0 is negligible, then these are simplified, as follows:

$$L_{WB} = \frac{l_T}{l}W \qquad L_T = \frac{l_{WB}}{l}W \tag{4.77}$$

where

$$\lambda_T = \frac{l_T}{l} \qquad \lambda_{WB} = \frac{l_{WB}}{l}$$

Aircraft drag can be written as the combination of wing/body drag and tail drag:

$$D = QS_W\left(C_{D0} + k_{D_{WB}}\left(\lambda_T\frac{W}{QS_W}\right)^2\right) + QS_T K_{D_T}\left(\lambda_{WB}\frac{W}{QS_T}\right)^2 \tag{4.78}$$

where C_{D0} is construed as the profile drag coefficient for the whole aircraft. This is rewritten as:

$$D = QS_W C_{D0} + \frac{W^2}{Q}F_D \tag{4.79}$$

where

$$F_D = (1 - \lambda_{WB})^2\frac{k_{D_{WB}}}{S_W} + \lambda_{WB}^2\frac{k_{D_T}}{S_T} \tag{4.80}$$

Differentiating with respect to Q:

$$\frac{dD}{dQ} = S_W C_{D0} - \frac{W^2}{Q^2} F_D$$

Minimum drag D_{min} and the corresponding dynamic pressure Q_{md} occur when $dD/dQ = 0$:

$$Q_{md} = \frac{W}{\sqrt{S_W C_{D0}}} \sqrt{F_D} \qquad D_{min} = 2W \sqrt{K_D C_{D0}} \sqrt{F_D} \qquad (4.81)$$

The optimal CG position for minimum drag is established by solving:

$$\frac{dD_{min}}{d\lambda_{WB}} = 2W \sqrt{K_D C_{D0}} \frac{1}{\sqrt{F_D}} \frac{dF_D}{d\lambda_{WB}} = 0$$

which is equivalent to solving:

$$0 = \frac{dF_D}{d\lambda_W} = -2(1 - \lambda_{WB}) \frac{k_{D_{WB}}}{S_W} + 2\lambda_{WB} \frac{k_{D_T}}{S_T}$$

$$\left(\frac{k_{D_{WB}}}{S_W} + \frac{k_{D_T}}{S_T} \right) \lambda_{WB} = \frac{k_{D_{WB}}}{S_W}$$

Thus, the optimal CG position is found:

$$\lambda_w = \frac{k_{D_{WB}} S_T}{k_{D_T} S_W + k_{D_{WB}} S_T} \qquad (4.82)$$

For this value of λ_{WB}, the function F_D becomes:

$$F_D = \frac{k_{D_{WB}} k_{D_T}}{k_{D_T} S_W + k_{D_{WB}} S_T} \qquad (4.83)$$

Back substitution into Equation 4.81 gives the values of Q_{md} and D_{min}.

The distance from the aircraft datum point to the minimum drag point is determined as l_D, where:

$$l_D = l \frac{k_{D_{WB}} S_T}{k_{D_T} S_W + k_{D_{WB}} S_T} \qquad (4.84)$$

Assuming that the induced drag factors are similar, this can be roughly approximated as:

$$l_D \approx l \frac{S_T}{S_W + S_T} \qquad (4.85)$$

Comparison with Equation 4.62 shows that the minimum possible drag is achieved when point D is roughly coincident with the neutral point N, i.e. drag minimisation requires neutral stability.

4.9 Steady-State Flight Performance

4.9.1 Definitions

Specific Air Range (SAR) is the distance travelled through air per unit of fuel burnt and Specific Endurance (SE) is the time elapsed per unit of fuel burnt.

Airspeed V and fuel flow F are defined as:

$$V = \frac{dx}{dt} \qquad F = -\frac{dm}{dt} \qquad (4.86)$$

Thus, the parameters of interest are determined as:

$$SAR = -\frac{dx}{dm} = -\frac{dx}{dt}\frac{dt}{dm} = \frac{V}{F} \qquad SE = -\frac{dt}{dm} = \frac{1}{F} \qquad (4.87)$$

Alternatively, SAR can be defined using referred airspeed V_0 [cf. Equation 1.18] and referred fuel flow F_0 [cf. Equation 1.21], which leads to the standard definition of SAR-δ:

$$SAR = \frac{V}{F_0\delta\sqrt{\theta}} = \frac{V_0}{F_0\delta} \qquad \Longrightarrow \qquad SAR - \delta = \frac{V_0}{F_0} \qquad (4.88)$$

When using referred parameters, it is good practice to quote airspeed as a Mach number, such that:

$$SAR - \delta = \frac{v_0}{F_0}M \qquad (4.89)$$

where v_0 is the speed of sound at sea level.

4.9.2 Airspeeds for Maximum Endurance and Maximum Range

For a thrust-producing engine, fuel flow is proportional to nett thrust. The constant of proportionality is the Specific Fuel Consumption (f), which is defined as the ratio of fuel flow F to nett thrust N:

$$f = \frac{F}{N} \qquad (4.90)$$

This number will vary with airspeed and altitude although, for level flight performance at constant airspeed, it will be constant. In this condition, specific endurance is inversely proportional to drag (noting that D equals nett thrust N):

$$SE = \frac{1}{F} = \frac{1}{fN} = \frac{1}{f}\frac{1}{D} \qquad (4.91)$$

Specific endurance is maximised and achieved when drag is minimised:

$$SE_{max} \propto \frac{1}{D_{min}} \qquad (4.92)$$

In other words, the best endurance speed for a jet aircraft is identically the minimum drag speed.

The SAR is proportional to airspeed divided by drag (V/D):

$$SAR = \frac{V}{F} = \frac{V}{fN} = \frac{V}{fD}$$

(4.93)

This is maximised when V/D is maximised or, equivalently, D/V is minimised. Aircraft drag is defined by:

$$D = QSC_{D0} + \frac{K_D}{QS}W^2$$

where $Q = \rho V^2/2$ is dynamic pressure. Therefore:

$$\frac{D}{V} = \frac{\rho V}{2}SC_{D0} + \frac{2}{\rho V^3}\frac{K_D}{S}W^2$$

Differentiate with respect to airspeed (V) and set the result to zero in order to find the airspeed that corresponds with minimum D/V:

$$\frac{d}{dV}\left(\frac{D}{V}\right) = \frac{\rho}{2}SC_{D0} - 3\frac{2}{\rho V^4}\frac{K_D}{S}W^2 = 0$$

This gives the maximum range speed V_{mr}:

$$\frac{1}{2}\rho V_{mr}^2 = \sqrt{3}\sqrt{\frac{K_D}{C_{D0}}}W$$

(4.94)

From Equation 4.67, the minimum drag speed V_{md} is given by:

$$\frac{1}{2}\rho V_{md}^2 = \frac{2}{\rho}\sqrt{\frac{K_D}{C_{D0}}}\left(\frac{W}{S_W}\right)$$

Thus, the relationship between minimum drag speed and maximum range speed is:

$$V_{mr}^2 = \sqrt{3}\,V_{md}^2$$

(4.95)

Alternatively:

$$V_{mr} = \sqrt[4]{3}\,V_{md} = 1.31607V_{md}$$

(4.96)

4.9.3 Range and Endurance

For level-flight equilibrium, lift equals weight. This applies along the full length of the cruise path, during which the weight of the aircraft will decrease as fuel is burned (from an initial weight W_i to a final weight W_f). Aerodynamic lift depends on three parameters, namely lift coefficient (which depends on AOA α), altitude (via density ρ), and airspeed (via airspeed V). One of these parameters must be sacrificed in order to vary the lift in response to the decrease in weight while the other two parameters are held constant. Thus, there are three standard cruise methods. The current discussion will consider the **cruise method**, which maintains constant altitude and constant airspeed. This implies that the lift coefficient must decrease as weight decreases and that thrust must decrease as drag decreases.

Using Equations 4.86 and 4.90, the instantaneous SAR can be written as:

$$\frac{dx}{dm} = -\frac{V}{fD} \tag{4.97}$$

where it is noted that drag D is equal to nett thrust N. Also, the Specific Fuel Consumption (SFC) is constant in this flight condition. An equivalent formulation is:

$$g\frac{dx}{dW} = -\frac{V}{fD} \tag{4.98}$$

where $W = mg$ (i.e. weight equal mass multiplied by gravitational acceleration).
    Equation 4.65 gives the aircraft trim drag as:

$$D = QSC_{D0} + \frac{K_D}{QS_W}W^2$$

This equation can be written in the form:

$$D = a\left(1 + (bW)^2\right) \tag{4.99}$$

where

$$a = QSC_{D0} \qquad b = \frac{1}{qS}\sqrt{\frac{K_D}{C_{D0}}} \qquad \Longrightarrow \qquad ab = \sqrt{K_DC_{D0}} \tag{4.100}$$

The Range Equation is obtained by integrating Equation 4.98:

$$\Delta x = x_f - x_i = -\frac{V}{gf}\int_{W_i}^{W_f}\frac{dW}{a(1 + bW^2)} = -\frac{V}{gf}\frac{1}{ab}\left[\tan^{-1}(bW)\right]_{W_i}^{W_f}$$

In its evaluated form:

$$\Delta x = \frac{V}{gf}\frac{1}{\sqrt{K_DC_{D0}}}\left[\tan^{-1}\left(\frac{W}{qS}\sqrt{\frac{K_D}{C_{D0}}}\right)\right]_{W_f}^{W_i} \tag{4.101}$$

Recalling Equations 4.87, it is clear that $SAR = V.SE$, where airspeed V is constant for this cruise method. Therefore, the Endurance Equation can be written:

$$\Delta t = t_f - t_i = \frac{1}{gf}\frac{1}{\sqrt{K_DC_{D0}}}\left[\tan^{-1}\left(\frac{W}{qS}\sqrt{\frac{K_D}{C_{D0}}}\right)\right]_{W_f}^{W_i} \tag{4.102}$$

4.9.4 Alternative Form for Jet Aircraft Range and Endurance

It is convenient to use the following nondimensional parameters:

$$w = \frac{W_i}{W_f} \qquad u = \frac{V}{V_{md}} = \sqrt{\frac{q}{q_{md}}} \qquad \lambda_{max} = \frac{L}{D_{min}} \tag{4.103}$$

The minimum drag condition was stated as:

$$D_{min} = 2W\sqrt{K_D C_{D0}} \qquad q_{md} = \sqrt{\frac{K_D}{C_{D0}}\left(\frac{W}{S}\right)}$$

The following deductions can be made:

$$\sqrt{K_D C_{D0}} = \frac{D_{min}}{2W} \qquad\qquad \frac{W}{qS}\sqrt{\frac{K_D}{C_{D0}}} = \frac{q_{md}}{q}$$

$$\frac{1}{\sqrt{K_D C_{D0}}} = 2\frac{L}{D_{min}} = 2\lambda_{max} \qquad \frac{W}{qS}\sqrt{\frac{K_D}{C_{D0}}} = \frac{q_{md}}{q} = \frac{1}{u^2}$$

In the **cruise method**, airspeed V remains constant. Weight decreases over time and so too does the minimum drag speed because $V_{md} \propto \sqrt{W}$. The relationship between initial and final values of relative speed is given by:

$$V = u_i V_{mdi} = u_f V_{mdf} \implies u_f = u_i\frac{V_{mdi}}{V_{mdf}} = u_i\sqrt{\frac{W_i}{W_f}} = u_i\sqrt{w} \qquad (4.104)$$

the Jet Aircraft Range Equation for a constant altitude, constant Mach cruise was given by Equation 4.101. In its alternative form, it is given by:

$$\Delta x = 2\left(\frac{\lambda_{max}}{gf}V_{mdi}\right)u_i\left[\tan^{-1}\left(\frac{1}{u_i^2}\right) - \tan^{-1}\left(\frac{1}{wu_i^2}\right)\right] \qquad (4.105)$$

By similar deduction, the corresponding Endurance Equation is:

$$\Delta t = 2\left(\frac{\lambda_{max}}{gf}\right)\left[\tan^{-1}\left(\frac{1}{u_i^2}\right) - \tan^{-1}\left(\frac{1}{wu_i^2}\right)\right] \qquad (4.106)$$

4.9.5 Fuel Required to Carry Fuel

The problem with chemical fuel is that extra fuel is needed in order to carry the fuel necessary to propel the aircraft. Hypothetically, if the fuel were weightless, then aircraft weight would be constant; lift coefficient, altitude, and airspeed would also remain constant. Therefore, there would be one and only one cruise method. Under this hypothesis, the cruise condition would be the 'final' cruise condition from the previous discussion. Using the same notation, the fuel weight required in order to achieve the same endurance Δt would be:

$$W_t = Q_f\,\Delta t = gfX_N\Delta t = gfD\,\Delta t$$

$$W_t = gf\left(qSC_{D0} + \frac{K_D}{qS}W_f^2\right)\Delta t \qquad (4.107)$$

The actual fuel consumed is equal to the initial weight W_i minus the final weight W_f. So, the penalty for having to carry fuel is given by:

$$\Delta W = \left(W_i - W_f\right) - W_t \qquad (4.108)$$

4.10   Dynamic Modes

No discussion of aircraft flight is complete without some consideration of flight dynamics and, in particular, dynamic modes. This chapter (and this book) is focused on longitudinal flight, driven by lift, drag, and pitching moment. Chapter 1 incorporated a rolling moment so that an aircraft could execute horizontals turns. The resulting model had four degrees of freedom, which had the virtue of being simple but had the drawback of missing yaw and sideslip. So, in effect, control of the idealised aircraft is based on coordinated turns and side-slip suppression. What this means for dynamic analysis is that longitudinal modes can be identified and characterised, plus a pure roll mode. Dutch roll mode and spiral mode are not identifiable because the necessary degrees of freedom are not present in the model.

Recalling Section 1.2, aircraft motion is defined by Equations 1.7–1.9:

$$\dot{V} = \frac{T-D}{m} - g\sin\gamma_2 \qquad \dot{\gamma}_2 = \frac{L}{mV}\cos\gamma_1 - \frac{g}{V}\cos\gamma_2 \qquad \dot{\gamma}_3 = \frac{L}{mV}\frac{\sin\gamma_1}{\cos\gamma_2}$$

$$\dot{p} = \frac{K}{J_1} \qquad \dot{\gamma}_1 = p \qquad \dot{q} = \frac{M}{J_2} \qquad \dot{\alpha} = q - \dot{\gamma}$$

$$\dot{e} = V\cos\gamma_2\sin\gamma_3 \qquad \dot{n} = V\cos\gamma_2\cos\gamma_3 \qquad \dot{h} = -V\sin\gamma_2$$

where m is the aircraft mass, g is the gravitational acceleration, V is the airspeed, J_1 is the roll moment of inertia, and J_2 is the pitch moment of inertia. The flight path angles are $(\gamma_1, \gamma_2, \gamma_3)$ and aircraft position is measured as $e =$ East, $n =$ North, and $h =$ Altitude. The force/moment system is shown in Figure 4.14, together with pitch rate q and roll rate p (which acts about the velocity vector).

In addition, Equation 1.6 tracks the variation in the flight path direction measured in the longitudinal plane of symmetry. This was given a separate parameter γ without a subscript.

$$\dot{\gamma} = \dot{\gamma}_2\cos\gamma_1 + \dot{\gamma}_3\cos\gamma_2\sin\gamma_1$$

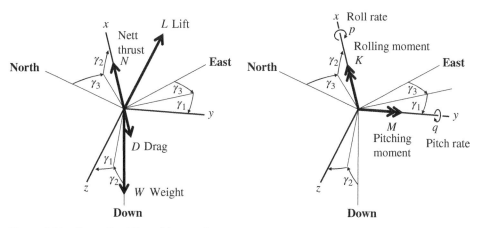

Figure 4.14   Generalised Force/Moment System.

The absolute value of γ is not important but its rate of change enables $\dot\alpha$ (and therefore α) to be calculated, as indicated above.

For wings-level flight, in which $\gamma_1 = 0$ and $\dot\gamma = \dot\gamma_2$, AOA is calculated as:

$$\dot\alpha = q - \dot\gamma \quad \Longrightarrow \quad \alpha = \theta - \gamma \tag{4.109}$$

where θ is the aircraft pitch angle (as implied by Figure 1.1). Setting aside the position updates, the remaining equations are simplified for straight-and-level flight ($\gamma_2 = 0$), as follows:

$$m\dot V = N - D \qquad mV(q - \dot\alpha) = L - W \qquad \dot\gamma_3 = 0 \tag{4.110}$$

$$J_1\dot p = K \qquad \dot\varphi = p \qquad J_2\dot q = M \qquad \dot\theta = q \tag{4.111}$$

where φ is the aircraft roll angle ($\varphi = \gamma_1$). The longitudinal equations of motion are summarised as:

$$\begin{pmatrix} m & 0 & 0 & 0 \\ 0 & mV & 0 & 0 \\ 0 & 0 & J_2 & 0 \\ 0 & 0 & 0 & 1 \end{pmatrix} \begin{pmatrix} \dot V \\ \dot\alpha \\ \dot q \\ \dot\theta \end{pmatrix} = \begin{pmatrix} 0 & 0 & 0 & 0 \\ 0 & 0 & mV & 0 \\ 0 & 0 & 0 & 0 \\ 0 & 0 & 1 & 0 \end{pmatrix} \begin{pmatrix} V \\ \alpha \\ q \\ \theta \end{pmatrix} + \begin{pmatrix} N - D \\ W - L \\ M \\ 0 \end{pmatrix} \tag{4.112}$$

The (one and only) lateral equation of motion is:

$$J_1\dot p = K \tag{4.113}$$

Linearisation is a process for representing the dynamic behaviour of the aircraft as it is perturbed from an equilibrium flight condition. Essentially, forces and moments are divided into linear components, each relating to one causal factor (e.g. AOA or tail rotation). It also removes any constants related to zero AOA or zero lift (most obviously C_{L0}, C_{D0}, and C_{M0}).

The lift acting on an aircraft can be written as:

$$L = QS_W L'_{WB} + QS_T L'_T$$

where $Q = \rho V^2/2$ is the dynamic pressure, S_W is the wing reference area, S_T is the tail reference area. Using the algebra from earlier sections, the respective lift coefficients[1] are L'_{WB} and L'_T:

$$L'_{WB} = a_{WB}(\alpha - \alpha_0) \qquad L'_T = a'_T\alpha + \frac{l_T}{V}a_Tq + a_T\delta_T + a_T(-\varepsilon_0 + \varepsilon_\alpha t'\dot\alpha + \varepsilon_\delta\delta_F)$$

Assuming that the flap deflection δ_F is fixed, these expressions are linearised, as follows:

$$\Delta L'_{WB} = a_{WB}\Delta\alpha \qquad \Delta L'_T = a'_T\Delta\alpha + \frac{l_T}{V}a_T\Delta q + a_T\Delta\delta_T + a_T\varepsilon_\alpha t'\Delta\dot\alpha$$

---

1 The notational L' to denote lift coefficient is borrowed from ship/submarine dynamics.

The lift force is linearised as:

$$\Delta L = L_V \, \Delta V + L_\alpha \, \Delta\alpha + L_q \, \Delta q + L_\delta \, \Delta\delta_T + L_{\dot\alpha} \, \Delta\dot\alpha \tag{4.114}$$

where

$$L_V = \rho V \left(S_W L'_{WB} + S_T L'_T \right) = \left(\frac{2}{V} \right) L$$

$$L_\alpha = QS_W a_{WB} + QS_T a'_T \qquad L_q = \frac{QS_T l_T a_T}{V} \qquad L_\delta = QS_T a_T \qquad L_{\dot\alpha} = QS_T a_T \varepsilon_\alpha t'$$

These quantities are 'derivatives' and they define the linear variations of lift when flying close to flight equilibrium.

Similarly, the pitching moment acting on an aircraft can be written as:

$$M = M_0 + QS_W l_{WB} L'_{WB} - QS_T l_T L'_T$$

Its linearised form is:

$$\Delta M = M_V \, \Delta V + M_\alpha \, \Delta\alpha + M_q \, \Delta q + M_\delta \, \Delta\delta_T + M_{\dot\alpha} \, \Delta\dot\alpha \tag{4.115}$$

where

$$M_V = (2/V)M \qquad M_\alpha = QS_W l_{WB} a_{WB} - QS_T l_T a'_T$$

$$M_q = \frac{-QS_T l_T^2 a_T}{V} \qquad M_\delta = -QS_T l_T a_T \qquad M_{\dot\alpha} = -QS_T l_T a_T \varepsilon_\alpha t'$$

The drag acting on an aircraft can be written as:

$$D = D_0 + QS_W k_{WB} \left(L'_{WB} \right)^2 + QS_T k_T \left(L'_T \right)^2$$

Its linearised form is:

$$\Delta D = 2Q \left(S_W k_{WB} L'_{WB} \Delta L'_{WB} + S_T k_T L'_T \Delta L'_T \right)$$

Algebraically, this is slightly less convenient to deal with but straightforward manipulation will reduce to the standard form:

$$\Delta D = D_V \Delta V + D_\alpha \Delta\alpha + D_q \Delta q + D_\delta \Delta\delta_T + D_{\dot\alpha} \Delta\dot\alpha \tag{4.116}$$

The linearised equations of longitudinal motion are considered as follows:

$$
\begin{pmatrix} m & D_{\dot\alpha} & 0 & 0 \\ 0 & mV + L_{\dot\alpha} & 0 & 0 \\ 0 & -M_{\dot\alpha} & J_2 & 0 \\ 0 & 0 & 0 & 1 \end{pmatrix}
\begin{pmatrix} \dot V \\ \dot\alpha \\ \dot q \\ \dot\theta \end{pmatrix}
=
\begin{pmatrix} -D_V & -D_\alpha & -D_q & -W \\ -L_V & -L_\alpha & -L_q + mV & 0 \\ M_V & M_\alpha & M_q & 0 \\ 0 & 0 & 1 & 0 \end{pmatrix}
\begin{pmatrix} V \\ \alpha \\ q \\ \theta \end{pmatrix}
+
\begin{pmatrix} 1 & -D_\delta \\ 0 & -L_\delta \\ 0 & M_\delta \\ 0 & 0 \end{pmatrix}
\begin{pmatrix} T \\ \delta_T \end{pmatrix}
$$

$$\tag{4.117}$$

The linearised equation of lateral of motion is:

$$J_1 \Delta \dot{p} = K_p \Delta p + K_\delta \Delta \delta_A \tag{4.118}$$

where the derivatives can be readily derived from the aerodynamics presented in Section 1.5.4.

Applying Laplace transforms:

$$\begin{pmatrix} sm + D_V & D_\alpha + D_{\dot{\alpha}} & D_q & W \\ L_V & s(mV + L_{\dot{\alpha}}) + L_\alpha & -L_q - mV & 0 \\ -M_V & -M_\alpha - M_{\dot{\alpha}} & sJ_2 - M_q & 0 \\ 0 & 0 & -1 & s \end{pmatrix} \begin{pmatrix} V \\ \alpha \\ q \\ \theta \end{pmatrix} = \begin{pmatrix} 1 & -D_{\delta_T} \\ 0 & -L_{\delta_T} \\ 0 & M_{\delta_T} \\ 0 & 0 \end{pmatrix} \begin{pmatrix} T \\ \delta_T \end{pmatrix} \tag{4.119}$$

and

$$(sJ_1 - K_p) \Delta p = K_{\delta_A} \Delta \delta_A \tag{4.120}$$

Dynamic modes are determined by the characteristic equations:

$$\Delta_{lon} = \begin{vmatrix} sm + D_V & D_\alpha + D_{\dot{\alpha}} & D_q & W \\ L_V & s(mV + L_{\dot{\alpha}}) + L_\alpha & -L_q - mV & 0 \\ -M_V & -M_\alpha - M_{\dot{\alpha}} & sJ_2 - M_q & 0 \\ 0 & 0 & -1 & s \end{vmatrix} = 0 \tag{4.121}$$

and

$$\Delta_{roll} = sJ_1 - K_p = 0 \tag{4.122}$$

Equation 4.122 defines the *roll mode*, which is a simple first-order lag, such that:

$$\Delta_{roll} = 1 + \tau_r s = 0 \tag{4.123}$$

which implies a time constant τ, where:

$$\tau_r = \frac{J_1}{(-K_p)} \tag{4.124}$$

The longitudinal dynamics divide into short period and long period modes. First of all, simplify Equation 4.121 by removing the explicit dependency on pitch angle θ:

$$\Delta_{lon} = \begin{vmatrix} sm + D_V & D_\alpha + D_{\dot{\alpha}} & D_q + W/s \\ L_V & s(mV + L_{\dot{\alpha}}) + L_\alpha & -L_q - mV \\ -M_V & -M_\alpha - M_{\dot{\alpha}} & sJ_2 - M_q \end{vmatrix} = 0 \tag{4.125}$$

Next, ignore the (small) contribution of $\dot{\alpha}$ in this context:

$$\Delta_{lon} = \begin{vmatrix} sm + D_V & D_\alpha & W/s \\ L_V & smV + L_\alpha & -mV \\ -M_V & -M_\alpha & sJ_2 - M_q \end{vmatrix} = 0 \tag{4.126}$$

Then, isolate the fast dynamics associated with the variation of pitching moment:

$$\Delta_{sp} = \begin{vmatrix} smV & -mV \\ -M_\alpha & sJ_2 - M_q \end{vmatrix} = 0 \tag{4.127}$$

This is expanded to give:

$$smV(sJ_2 - M_q) - mVM_\alpha = 0$$

$$s^2 - s\frac{M_q}{J_2} - \frac{M_\alpha}{J_2} = 0$$

This is the *short period mode*, which has a second-order response, such that:

$$\Delta_{sp} = s^2 - 2\zeta_{sp}\omega_{sp}s - \omega_{sp} = s^2 - s\frac{M_q}{J_2} - \frac{M_\alpha}{J_2} = 0 \tag{4.128}$$

where the frequency and damping ratio are:

$$\omega_{sp} = \sqrt{\frac{(-M_\alpha)}{J_2}} \qquad \zeta_{sp} = \frac{1}{2}\frac{(-M_q)}{\sqrt{J_2(-M_\alpha)}} \tag{4.129}$$

Now, isolate the slow dynamics that are associated with airspeed and gravity and, also, remove the 'fast' rates of change driven by smV and sJ_2:

$$\begin{vmatrix} sm + D_V & D_\alpha & W/s \\ L_V & smV + L_\alpha & -mV \\ -M_V & -M_\alpha & sJ_2 - M_q \end{vmatrix} \implies \begin{vmatrix} sm + D_V & 0 & W/s \\ L_V & L_\alpha & -mV \\ 0 & -M_\alpha & -M_q \end{vmatrix}$$

The characteristic equation becomes:

$$\Delta_{lp} = \begin{vmatrix} sm + D_V & 0 & W/s \\ L_V & L_\alpha & -mV \\ 0 & -M_\alpha & -M_q \end{vmatrix} = 0 \tag{4.130}$$

This is expanded using Schur's formula [Gantmacher (1959)]:

$$\Delta_{lp} \approx sm + D_V - \begin{pmatrix} 0 & W/s \end{pmatrix}\begin{pmatrix} L_\alpha & -mV \\ -M_\alpha & -M_q \end{pmatrix}^{-1}\begin{pmatrix} L_V \\ 0 \end{pmatrix} = 0$$

$$\Delta_{lp} \approx sm + D_V + \frac{L_V M_\alpha W/s}{L_\alpha M_q - mV M_\alpha} = 0$$

Finally, remove the dependency on (fast acting) M_q in order to give:

$$sm + D_V - \frac{L_V g/s}{V} = 0 \implies s^2 + \frac{D_V}{m}s - \frac{L_V}{m}\frac{g}{V} = 0$$

This is the *long period mode* (also known as the *phugoid*), which has a second-order response:

$$\Delta_{lp} \approx s^2 - 2\zeta_{sp}\omega_{sp}s - \omega_{sp} = s^2 + \frac{D_V}{m}s - \frac{L_V}{m}\frac{g}{V} \tag{4.131}$$

where the frequency and damping ratio are:

$$\omega_{lp} = \sqrt{2}\,\frac{g}{V} \qquad \zeta_{lp} = \frac{1}{\sqrt{2}}\frac{D_V}{L_V} \tag{4.132}$$

A final observation is on the variation of modal properties with airspeed and altitude. For the long-period mode, the damping ratio is invariant and the natural frequency $\omega_{lp} \propto 1/V$. For the short-period mode, the variation in relevant derivatives is $M_\alpha \propto Q$ and $M_q \propto Q/V$, where $Q = \rho V^2/2$. Thus, natural frequency and damping ratio vary as follows:

$$\omega_{sp} = \sqrt{\frac{(-M_\alpha)}{J_2}} \propto \sqrt{Q} \qquad \zeta_{sp} = \frac{1}{2}\frac{(-M_q)}{\sqrt{J_2(-M_\alpha)}} \propto \frac{\sqrt{Q}}{V}$$

Equivalently:

$$\omega_{sp} \propto V\sqrt{\rho} \qquad \zeta_{sp} \propto \sqrt{\rho} \tag{4.133}$$

5

Gas Turbine Dynamics

5.1 Introduction

5.1.1 The Importance of Gas Turbines

Gas turbines are everywhere! They are compact powerplants with a very high power-to-weight ratio and, more importantly, a very high power-to-volume ratio. They propel just about every air vehicle bigger than a light aircraft, either by generation of thrust or by generation of shaft power to drive a propeller or rotor. Therefore, the capability to model aircraft of significant size implies the capability to model gas turbines. Accordingly, a chapter is devoted to this subject in this book.

5.1.2 What Chapter 5 Includes

Chapter 5 includes:

- Ideal Gas Properties
- Gas Dynamics
- Engine Components
- Engine Dynamics
- Engine Models
- Gas Properties Data

5.1.3 What Chapter 5 Excludes

Chapter 5 excludes:

- A lot of detailed explanation of gas turbine technology and performance.
- Discussion of engine configurations apart from the basic distinction between turbojet and turbofan.
- Actual engine data (e.g. compressor and turbine maps).

5.1.4 Overall Aim

Chapter 5 should provide a starting point for establishing a gas turbine transient performance model, together with the calculations to estimate component behaviour. Essentially what is being provided is a framework for investigating the effect of engine parameters on achievable performance. The eventual outputs will only be as accurate as the engine data that populates the model but, even for preliminary or indicative data, the model is intended to produce the correct trends in performance.

5.2 Ideal Gas Properties

5.2.1 Equation of State

The equation of state relates pressure (P), temperature (T), and volume (V) of a gas. This comes from the gas laws of Boyle ($P \propto V^{-1}$), Charles ($T \propto V$), and Gay–Lussac ($P \propto T$). These merge into a combined law:

$$PV \propto T \tag{5.1}$$

At the microscopic scale, this can be written as:

$$PV = Nk_BT \tag{5.2}$$

where N = number of molecules of gas and $k_B = 1.3806488 \times 10^{-23}$ J. K^{-1} is the Boltzmann Constant. The amount of gas is given by the number of moles [n], where each mole contains $6.022140857 \times 10^{23}$ molecules [the Avogadro Number]. Thus, this equation can be written as:

$$PV = nN_Ak_BT = nR_0T \tag{5.3}$$

where $R_0 = N_Ak_B = 8.3144598$ J. mol^{-1}. K^{-1} is the universal gas constant. Equivalently:

$$PV = mRT \tag{5.4}$$

where m = mass of gas and R is the specific gas constant for a particular gas, such that $mR = nR_0$.

This produces the Ideal Gas Law, which is usually written as:

$$P = \rho RT \tag{5.5}$$

where the gas density ρ and the gas constant R are given by:

$$\rho = \frac{m}{V} \tag{5.6}$$

$$R = \frac{n}{m}R_0 = \frac{R_0}{\mu} \tag{5.7}$$

where μ is called the mean molecular mass.

5.2.2 Energy, Enthalpy, and Entropy

Consider energy transfers to and from a gas. For a given amount of gas, the *usable* energy supplied (ΔQ) is equal to the increase of internal energy (ΔU) as the gas heats up plus the work done by the gas (ΔW) by increasing its volume against the surrounding pressure. Conversely, the energy removed is equal to the decrease in internal energy plus the work done on the gas.

The Energy Equation is given by the *First Law of Thermodynamics*:

$$\Delta Q = \Delta U + \Delta W \tag{5.8}$$

In the special case where $\Delta Q = 0$, then $\Delta U = -\Delta W$. This defines an *Adiabatic Process*, in which no energy is exchanged between the process and its environment. This implies a process with very fast dynamics or with very small energy losses. The mechanical work is stated as $\Delta W = P\Delta V$, where ΔV is the volume change and P is the pressure that drives or opposes the volume change (depending whether volume is decreasing or increasing, respectively). In other words, the work done is equal to the change in stored energy, PV, for a given pressure P. So, the Energy Equation becomes:

$$\Delta Q = \Delta U + P\Delta V \tag{5.9}$$

Enthalpy, H, is the sum of Internal Energy and Pressure Energy:

$$H = U + PV \tag{5.10}$$

The enthalpy change associated with a process is:

$$\Delta H = \Delta U + P\Delta V + V\Delta P \tag{5.11}$$

Accordingly, the Energy Equation can be expressed in terms of internal energy or enthalpy via a combination of Equations 5.9 and 5.11:

$$\Delta Q = \Delta U + P\Delta V = \Delta H - V\Delta P \tag{5.12}$$

In real thermodynamic processes, energy is always dissipated in some way other than internal energy or useful work and, thus, the following inequality applies:

$$\Delta Q < T\Delta S$$

where ΔQ is the change in usable energy, ΔS is the change in entropy state, and T is temperature. Any attempt to reverse a process could never recover all the energy supplied and the initial state could never be re-established. Therefore, *real processes are irreversible*.

The *Second Law of Thermodynamics* states that:

$$\Delta Q \leq T\Delta S \tag{5.13}$$

It is convenient to postulate a *reversible process* that has no unaccounted losses, such that energy would be recovered fully if it were to be reversed. This could be achieved if and only if

$$\Delta S = 0 \qquad \Longrightarrow \qquad \Delta Q = 0$$

An ideal process has constant entropy; it is isentropic and therefore it must be adiabatic (reversible). Note that $\Delta Q = 0$ does not imply $\Delta S = 0$. So, an adiabatic process is not necessarily isentropic.

5.2.3 Specific Heat Capacity

Equation 5.12 gives rise to the following definitions for a mass of gas, m, undergoing a temperature change, ΔT. At constant volume, the usable energy supplied is equal to the change in internal energy:

$$\Delta V = 0 \quad \Longrightarrow \quad \Delta Q = \Delta U \triangleq mC_V\Delta T \tag{5.14}$$

At constant pressure, the usable energy supplied is equal to the change in enthalpy:

$$\Delta P = 0 \quad \Longrightarrow \quad \Delta Q = \Delta H \triangleq mC_P\Delta T \tag{5.15}$$

The *Specific Heat Capacity* of the gas is denoted by C_V at *constant volume* and denoted by C_P at *constant pressure*.

The differential relationship between enthalpy and internal energy is derived from Equations 5.11, 5.14, and 5.15:

$$P\Delta V + V\Delta P = \Delta H - \Delta U = m(C_P - C_V)\Delta T$$

The Ideal Gas Law [Equation 5.4] implies that:

$$P\Delta V + V\Delta P = mR\Delta T$$

Combining these relationships gives Mayer's Law:

$$R = C_P - C_V \tag{5.16}$$

5.2.4 Adiabatic Gas Ratio

The *Adiabatic Gas Ratio* is the ratio of specific heat capacities and is defined by:

$$\gamma = C_P/C_V \tag{5.17}$$

At constant volume, heating a gas will increase the internal energy. At constant pressure, internal energy will increase plus work is done against an opposing pressure. Hence, $\gamma > 1$. The actual value of γ depends on the number of degrees of freedom of the molecule, f (i.e. translation, rotation, and vibration). Each represents an energy store, in accordance with the kinetic theory of gases. Ideally, it is expected that $\gamma = 1 + 2/f$.

Note	Monatomic molecules have three translational degrees of freedom (γ = 5/3). *Example values are He:1.66, Ne:1.64, and Ar:1.67.*
	Diatomic molecules have three translational and two rotational degrees of freedom (γ = 7/5). *Example values are N_2:1.40, O_2:1.40, Dry Air:1.40, and H_2:1.41.*
	Triatomic molecules have six degrees of freedom (γ = 8/6). *Example values are H_2O:1.33, NH_2:1.32, and CO_2:1.28.*

5.2.5 Compressible Gas Properties

Now, consider the relationship between pressure, temperature, and volume of a compressible gas. From Equation 5.11:

$$\Delta Q = \Delta H - V \Delta P$$

where enthalpy H is associated with the specific heat capacity at constant pressure [cf. Equation 5.13] and volume V obeys the Ideal Gas Law.

For an adiabatic process, $\Delta Q = 0$:

$$\Delta H = V \Delta P \qquad \Longrightarrow \qquad mC_P \Delta T \equiv \frac{mRT}{P} \Delta P \qquad \Longrightarrow \qquad C_P \frac{\Delta T}{T} = R \frac{\Delta P}{P}$$

Integrating this equation between two gas conditions (denoted by subscripts '1' and '2'):

$$C_P \ln\left(\frac{T_2}{T_1}\right) = R \ln\left(\frac{P_2}{P_1}\right)$$

Applying Mayer's Law:

$$\ln\left(\frac{T_2}{T_1}\right) = \frac{\gamma - 1}{\gamma} \ln\left(\frac{P_2}{P_1}\right) \tag{5.18}$$

Thus, the adiabatic pressure–temperature relationship is established as:

$$\frac{T_2}{T_1} = \left(\frac{P_2}{P_1}\right)^{\frac{\gamma-1}{\gamma}} \qquad \text{or} \qquad \frac{P_2}{P_1} = \left(\frac{T_2}{T_1}\right)^{\frac{\gamma}{\gamma-1}} \tag{5.19}$$

Also, the Ideal Gas Law provides the following link between gas states:

$$\frac{P_2 V_2}{P_1 V_1} = \frac{T_2}{T_1} \qquad \Longrightarrow \qquad \ln\left(\frac{P_2}{P_1}\right) + \ln\left(\frac{V_2}{V_1}\right) = \ln\left(\frac{T_2}{T_1}\right)$$

Thus, Equation 5.19 becomes:

$$\ln\left(\frac{P_2}{P_1}\right) + \ln\left(\frac{V_2}{V_1}\right) = \frac{\gamma - 1}{\gamma} \ln\left(\frac{P_2}{P_1}\right) \qquad \Longrightarrow \qquad \frac{1}{\gamma} \ln\left(\frac{P_2}{P_1}\right) + \ln\left(\frac{V_2}{V_1}\right) = 0$$

This gives the Adiabatic Gas Law:

$$PV^\gamma = \text{constant} \qquad \text{or} \qquad \frac{P}{\rho^\gamma} = \text{constant} \tag{5.20}$$

5.2.6 Polytropic Processes

The Adiabatic Gas Law can be generalised in the following way:

$$PV^n = \text{constant} \tag{5.21}$$

In this context, the parameter n is called the *polytropic index* and it replaces the adiabatic gas ratio in order to express the full range of thermodynamic processes.

Table 5.1 Thermodynamic Processes.

Polytropic Index	Relationship	Process
$n = 0$	$V_1^{-1} T_1 = V_2^{-1} T_2 \qquad P_1 = P_2$	Isobaric
$n = 1$	$P_1 V_1 = P_2 V_2 \qquad T_1 = T_2$	Isothermal
$n < \gamma$		Polytropic expansion
$n = \gamma$	$\dfrac{T_{02}}{T_{01}} = \left(\dfrac{P_{02}}{P_{01}}\right)^{\frac{n-1}{n}}$	Isometric or adiabatic
$n > \gamma$		Polytropic compression
$n \to \infty$	$T_1 P_1^{-1} = T_2 P_2^{-1} \qquad V_1 = V_2$	Isometric

The adiabatic pressure–temperature relationship [Equation 5.19] can be rewritten as:

$$\frac{T_2}{T_1} = \left(\frac{P_2}{P_1}\right)^{\frac{n-1}{n}} \qquad \text{or} \qquad \frac{P_2}{P_1} = \left(\frac{T_2}{T_1}\right)^{\frac{n}{n-1}} \tag{5.22}$$

This can be interpreted as follows:

$$\frac{T_2}{T_1} = \left(\frac{P_2}{P_1}\right)^{\frac{n-1}{n}} \qquad \text{or} \qquad \frac{P_2}{P_1} = \left(\frac{T_2}{T_1}\right)^{\frac{n}{n-1}} \tag{5.23}$$

Temperature will increase more than for isentropic compression because extra work has to be done on the gas to achieve a given pressure ratio. Conversely, temperature will decrease less than for isentropic expansion because a given pressure ratio can be achieved with less work done by the gas. A summary of processes and their associated polytropic indices is presented in Table 5.1.

5.3 Gas Dynamics

5.3.1 Fundamental Relationships for Gas Flow

Consider a small volume of gas moving with a bulk velocity, $\vec{U}$, as shown on the left of Figure 5.1. For a given density, ρ, length, Δx, and cross-sectional area, A, it has mass:

$$\Delta m = \rho A \Delta x \tag{5.24}$$

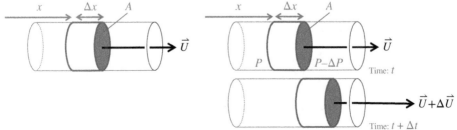

Figure 5.1 Massflow Through a Streamtube.

Without loss of generality, this has been sketched as a cylindrical element moving along a straight path. Its velocity (measured over a short time period, Δt) is given by:

$$\vec{U} = \frac{\Delta x}{\Delta t} \tag{5.25}$$

Thus, its rate of flow (i.e. the massflow) is defined by the passage of mass across a plane (perpendicular the flow direction) per unit time:

$$\frac{\Delta m}{\Delta t} = \frac{\Delta m}{\Delta x} \frac{\Delta x}{\Delta t} = \rho A \vec{U} \tag{5.26}$$

Regardless of changes in density, area, or velocity, this quantity must remain constant in order for mass to be conserved. This gives the Massflow Equation:

$$\dot{m} = \rho A \vec{U} = \text{constant} \tag{5.27}$$

Now, consider a volume of gas moving with a bulk velocity, $\vec{U} = dx/dt$, that increases across a pressure drop, ΔP, as shown on the right of Figure 5.1. Its mass is $\Delta m = \rho A \, \Delta x$. The driving force is the product of the pressure drop and the cross-sectional area, A.

Using Newton's Second Law of Motion:

$$-A \Delta P = \Delta m \frac{d\vec{U}}{dt} \quad \Longrightarrow \quad A \frac{dP}{dx} \Delta x = -(\rho A \Delta x) \frac{d\vec{U}}{dx} \frac{dx}{dt} \quad \Longrightarrow \quad \frac{dP}{dx} = -\rho \vec{U} \frac{d\vec{U}}{dx}$$

This gives the Momentum Equation:

$$dP = -\rho \vec{U} \, d\vec{U} \tag{5.28}$$

5.3.2   Speed of Sound

Disturbances caused by an object within a compressible gas will propagate through the gas at the sound of sound (*approximately 340 m s^{-1} at sea level for dry air*). For a fixed object, a disturbance will radiate outwards in all directions. In one second, the original disturbance will be detectable at a range of 340 m because the resulting pressure wave will have travelled that far (shown in Figure 5.2). Successive pressure waves appear in a concentric pattern (in the left-hand sketch).

For a moving object, a disturbance will radiate outwards from instantaneous positions as the object passes through the gas. The object keeps moving on, leaving the pressure waves to propagate independently. The pattern is skewed in the direction of travel of the object. The relative speed of the advancing portion of the wave is less than that for the receding portion. As the object accelerates, the relative speed with respect to the advancing portion decreases until it becomes zero (in the centre sketch). The object is now moving at the speed of sound.

The *mechanical process* that underlies changes in pressure and density is constrained by the Adiabatic Gas Law [Equation 5.20]. This implies the following relationship:

$$P = k\rho^{\gamma} \tag{5.29}$$

where k is an arbitrary constant.

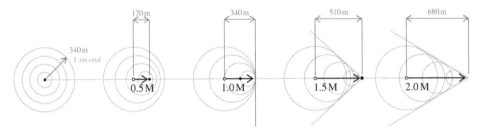

Figure 5.2 Sound Propagation Through Dry Air at Sea Level.

The flow process that underlies changes in pressure and density is constrained by Mass-flow [Equation 5.27] and Momentum [Equation 5.28]:

$$\rho A \vec{U} = \text{constant} \qquad\qquad dP = -\rho \vec{U}\,d\vec{U}$$

For flow through a constant cross-sectional area:

$$\rho \Delta \vec{U} + \vec{U}\Delta\rho = 0 \qquad \text{and} \qquad dP = \left(-\rho\,d\vec{U}\right)\vec{U} = \left(\vec{U}d\rho\right)\vec{U}$$

The speed of sound is established when the mechanical process opposes the flow process, i.e. when both processes yield the same rate of change of pressure relative to density:

$$\frac{dP}{d\rho} = k\gamma\rho^{\gamma-1} = (k\rho^{\gamma})\frac{\gamma}{\rho} = \gamma\frac{P}{\rho} \qquad\qquad \frac{dP}{d\rho} = \vec{U}^2 \quad\Longrightarrow\quad \vec{U}^2 = \gamma\frac{P}{\rho}$$

Applying the Ideal Gas Law, the speed of sound ($V \triangleq a$) is obtained in its familiar form:

$$a = \sqrt{\gamma R T} \tag{5.30}$$

Airspeeds can be quoted with reference to the local speed of sound. In this context, speed is expressed as a Mach Number, M, which is defined by:

$$M = \frac{\vec{U}}{a} \tag{5.31}$$

For dry air ($\gamma = 1.4$ and $R = 287.05287$) at sea level (where $T = 288.15$ K), the speed of sound is calculated as 340.29 m.s^{-1}. In Figure 5.2, Mach numbers of 0.5M, 1.0M, 1.5M, and 2.0M correspond approximately with speeds of 170, 340, 510, and 680 m.s^{-1}.

5.3.3 Bernoulli's Equation

Conservation of mechanical energy is established by integrating the momentum equation:

$$dP = -\rho \vec{U}\,d\vec{U}$$

Bernoulli's Equation for incompressible flow is obtained by integration, as follows:

$$\frac{P}{\rho} + \frac{\vec{U}^2}{2} = \text{constant} \tag{5.32}$$

Applying Equation 5.29, the Momentum Equation becomes:

$$\left(\frac{k}{P}\right)^{\frac{1}{\gamma}} dP = -\vec{U}d\vec{U} \qquad \Longrightarrow \qquad \frac{\gamma}{\gamma-1}\left(\frac{k}{P}\right)^{\frac{1}{\gamma}} P = -\frac{\vec{U}^2}{2} + \text{constant}$$

Bernoulli's Equation for compressible flow is then determined as follows:

$$\frac{\gamma}{\gamma-1}\left(\frac{P}{\rho}\right) + \frac{\vec{U}^2}{2} = \text{constant} \tag{5.33}$$

Using the Ideal Gas Law, the compressible form of Bernoulli's Equation is derived as:

$$\frac{C_P}{R}\left(\frac{P}{\rho}\right) + \frac{\vec{U}^2}{2} = \frac{C_P}{R}\left(\frac{\rho RT}{\rho}\right) + \frac{\vec{U}^2}{2} = \text{constant}$$

$$C_P T + \frac{\vec{U}^2}{2} = \text{constant} \tag{5.34}$$

5.3.4 Stagnation Conditions

Bernoulli's Equation for compressible flow leads directly to an extended definition of temperature that incorporates the kinetic energy of the gas flow, as well as the kinetic energy of molecular activity (as internal energy). This postulates an equivalent temperature measurement for a gas flow that is brought instantaneously to rest. This is established as:

$$C_P T + \frac{\vec{U}^2}{2} = C_P T_0$$

where T_0 is the stagnation temperature (or *total temperature*).

$$T_0 = T + \frac{\vec{U}^2}{2C_P} \tag{5.35}$$

Thus, Equation 5.34 can be written in simple form:

$$C_P T_0 = \text{constant} \tag{5.36}$$

The ratio of stagnation temperature, T_0, to what now has to be called *static temperature*, T, is:

$$\frac{T_0}{T} = 1 + \frac{\vec{U}^2}{2C_P T} \tag{5.37}$$

Using the definition of Mach number, together with Mayer's Law and the Adiabatic Gas Ratio, this ratio is obtained in its widely-used form:

$$\frac{T_0}{T} = 1 + \frac{M^2 \gamma RT}{2C_P T} = 1 + \frac{M^2 \gamma}{2}\left(\frac{C_P - C_V}{C_P}\right) = 1 + \frac{M^2 \gamma}{2}\left(\frac{\gamma-1}{\gamma}\right)$$

$$\frac{T_0}{T} = 1 + \frac{\gamma-1}{2}M^2 \tag{5.38}$$

The concept of stagnation temperature is an ideal snapshot of the state of a flow process. Flow is being treated as being brought to rest instantaneously (i.e. the transition is adiabatic) and static and stagnation temperatures are treated as interchangeable (i.e. the transition is reversible).

Thus, the transition is *isentropic* and can be expressed using Equation 5.19:

$$\frac{P_0}{P} = \left(\frac{T_0}{T}\right)^{\frac{\gamma}{\gamma-1}} \tag{5.39}$$

where P_0 = stagnation pressure and P = static pressure. Using Equation 5.38:

$$\frac{P_0}{P} = \left(1 + \frac{\gamma-1}{2}M^2\right)^{\frac{\gamma}{\gamma-1}} \tag{5.40}$$

Recall that Equation 5.19 relates static pressure and temperature ratios between two gas conditions (denoted by subscripts '1' and '2'). Equation 5.39 provides the transition between static and stagnation conditions. These are combined to show:

$$\left(\frac{P_{02}}{P_{01}}\right)\left(\frac{P_1}{P_2}\right) = \left(\frac{P_{02}}{P_2}\right)\left(\frac{P_1}{P_{01}}\right) = \left[\left(\frac{T_{02}}{T_2}\right)\left(\frac{T_1}{T_{01}}\right)\right]^{\frac{\gamma}{\gamma-1}}$$
$$= \left[\left(\frac{T_{02}}{T_{01}}\right)\left(\frac{T_1}{T_2}\right)\right]^{\frac{\gamma}{\gamma-1}} = \left(\frac{T_{02}}{T_{01}}\right)^{\frac{\gamma}{\gamma-1}}\left(\frac{P_1}{P_2}\right)$$

Thus, the relationship between stagnation pressure and temperature ratios between two gas conditions is also given by Equation 5.19:

$$\frac{P_{02}}{P_{01}} = \left(\frac{T_{02}}{T_{01}}\right)^{\frac{\gamma}{\gamma-1}} \tag{5.41}$$

An efficiency factor is incorporated for real gas processes, such that this equation is rewritten as:

$$\frac{P_{02}}{P_{01}} = \left(\frac{T_{02}}{T_{01}}\right)^{\frac{n}{n-1}} \tag{5.42}$$

where n is a polytropic index, as discussed in Section 5.2.6.

5.4 Engine Components

Gas turbine models will be discussed in Section 5.4. It is appropriate to define major components in advance so that the building blocks are fully presented and properly explained. Essentially, an engine requires a compressor, a combustor, a turbine, and a nozzle. In addition, there are ducts (which cause pressure losses) and junctions (where flow is split or flows are merged). A brief summary of gas properties is given in Section 5.6.

5.4.1 Duct

Following Bernoulli's description of internal flow, the relationship between the pressure drop across a duct and the flow through the duct is expressed as:

$$\Delta P = P_{01} - P_{02} = k \left(\frac{1}{2} \rho V^2 \right) \tag{5.43}$$

where P = stagnation pressure, ρ = gas density, V = flow velocity and k is a constant. The subscripts '1' and '2' denote inlet and outlet conditions, respectively. Note the additional subscript '0' denotes stagnation conditions.

A duct does not store mass or energy and so massflow $\dot{m}$ is conserved (which is re-expressed as $W_{in} = W_{out}$) and temperature is constant ($T_{in} = T_{out}$). Thus, this equation can be rewritten as:

$$\Delta P = P_{01} - P_{02} = \frac{k}{2\rho} \left(\frac{W_1}{A} \right)^2 \tag{5.44}$$

The average pressure in the duct can be related to the ideal gas law:

$$P = \frac{P_{01} + P_{02}}{2} = \rho R T_{01} \tag{5.45}$$

The pressure terms can be combined as:

$$(P_{01} - P_{02})(P_{01} + P_{02}) = \Delta P(P_{01} + (P_{01} - \Delta P)) = \Delta P(2P_{01} - \Delta P)$$

Thus, Equations 5.44 and 5.45 can be combined as:

$$\Delta P(2P_{01} - \Delta P) = k R T_{01} \left(\frac{W_1}{A} \right)^2$$

Assuming that the pressure drop is small, this can be approximated as:

$$2P_{01} \Delta P \approx \left(\frac{k}{A^2} \right) R T_{01} W_1^2 \tag{5.46}$$

This is rearranged to give the standard equation for a duct:

$$\frac{\Delta P}{P_{01}} = F_{PL} Q_1^2 \tag{5.47}$$

where F_{PL} is the *pressure loss factor* and where Q_1 is the flow function defined by:

$$Q_1 = \frac{W_1 \sqrt{T_{01}}}{P_{01}} \tag{5.48}$$

Accordingly, the outlet pressure is calculated as:

$$P_{02} = P_{01} \left(1 - F_{PL} Q_1^2 \right) \tag{5.49}$$

5.4.2 Junction

The splitting of one flow path into several new flow paths is trivial from a modelling perspective because the only parameter that is affected is massflow W. If the inlet conditions are labelled '1' and one of N outlets is labelled 'n', then the propagation is written simply as:

$$W_n = \lambda_n W_1 \qquad P_{0n} = P_{01} \qquad T_{0n} = T_{01} \tag{5.50}$$

where λ_n specifies the fraction of inlet flow that goes to output n. These factors are constrained by:

$$\sum_{n=1}^{N} \lambda_n = 1 \tag{5.51}$$

The merging of many flow paths into one requires addition of massflow W and energy flow Wh, where h is specific enthalpy. If outlet conditions are labelled '2' and inlets are labelled 'n', then:

$$W_2 = \sum_{n=1}^{N} W_n \qquad W_2 h_{02} = \sum_{n=1}^{N} W_n h_{0n} \tag{5.52}$$

The output temperature is derived from specific enthalpy (h_{02}), as discussed in Section 5.6.

5.4.3 Compressor

Generically, a compressor is a rotor that draws air from a low-pressure source and propels it into a higher-pressure volume downstream. A multi-stage compressor has multiple rotors that can achieve a much larger pressure at the outlet. Performance is defined by a *compressor map* that shows lines for pressure ratio P_{02}/P_{01} versus inlet flow function Q_1 over a range of aerodynamic speed N_{aero}:

$$N_{aero} = \frac{N}{\sqrt{T_{01}}} \tag{5.53}$$

where N is the actual rotational speed and T_{01} is the inlet stagnation temperature. An example is given in Figure 5.3. A separate graph shows pressure ratio versus efficiency.

For polytropic efficiency η_p, the temperature ratio T_{02}/T_{01} is calculated using Equation 5.42:

$$\frac{T_{02}}{T_{01}} = = \left(\frac{P_{02}}{P_{01}}\right)^{\frac{n-1}{n}} \tag{5.54}$$

where the polytropic compression index n is defined by:

$$\frac{n-1}{n} = \frac{1}{\eta_p}\left(\frac{\gamma-1}{\gamma}\right) \tag{5.55}$$

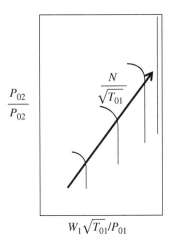

Figure 5.3 Example Compressor Map.

For isentropic efficiency η_i, the temperature ratio is:

$$\frac{T_{02}}{T_{01}} = 1 + \frac{1}{\eta_i}\left(\left(\frac{P_{02}}{P_{01}}\right)^{\frac{\gamma-1}{\gamma}} - 1\right) \tag{5.56}$$

The fact that the efficiency is less than one means that frictional losses cause the air to heat up more than would be expected for isentropic flow. Note that polytropic efficiency *is greater than* isentropic efficiency for compression.

The power consumed by a compressor is equal to the rate at which thermal energy is added to the massflow. The change in temperature can be converted into a change in specific enthalpy [cf. Section 5.6]. Therefore, for an inlet (and outlet) massflow W_1 that is heated from T_{01} to T_{02}, the required power (otherwise known as 'compressor work') is calculated as:

$$J_{compressor} = W_1(h_{02} - h_{01}) \tag{5.57}$$

Air bleed provides flow for environmental conditioning on the aircraft and for turbine cooling/sealing within the engine. Offtakes can be drawn from an intermediate compressor stage, labelled '12', necessitating a change to the calculations above. The simplest method is to take the pressure ratio after m out of M stages and use polytropic compression, as in Equation 5.54:

$$\frac{P_{012}}{P_{01}} = \left(\frac{P_{02}}{P_{01}}\right)^{\frac{m}{M}} \qquad \frac{T_{012}}{T_{01}} = = \left(\frac{P_{03}}{P_{01}}\right)^{\left(\frac{n-1}{n}\right)} \tag{5.58}$$

Thus, compressor power is re-calculated as:

$$J_{compressor} = W_1(h_{012} - h_{01}) + (W_{01} - W_{12})(h_{02} - h_{012})$$
$$J_{compressor} = W_1(h_{02} - h_{01}) - W_{12}(h_{02} - h_{012}) \tag{5.59}$$

where W_{12} is the bleed massflow. The resultant massflow at the compressor outlet is:

$$W_2 = W_1 - W_{12} \tag{5.60}$$

5.4.4 Split Compressor

A compressor can deliver flow into two concentric ducts, as is the case for turbofan engines. Flow can be modelled as passing through an *inner* compressor and an *outer* compressor, having separate compressor maps. Thus, the fan is modelled as a split compressor, with different pressure ratios and compression efficiencies between the inner and outer parts. The temperature ratio for the outer compressor will be greater than for the inner compressor because it operates at lower efficiency by virtue of high velocity at the blade tips. In this case, the compressor power is calculated:

$$J_{compressor} = W_2(h_{02} - h_{01}) + W_{21}(h_{021} - h_{01}) \tag{5.61}$$

where the inlet is labelled '1', the inner compressor is labelled '2', and the outer compressor is labelled '21'. The inlet massflow is determined by the split outlet conditions:

$$W_1 = W_2 + W_{21} \tag{5.62}$$

The *bypass ratio* is defined by:

$$\lambda = \frac{W_{21}}{W_2} \tag{5.63}$$

5.4.5 Combustor

Combustion adds a large amount of heat to the gas, resulting in a temperature rise:

$$\Delta T = T_{02} - T_{01} \tag{5.64}$$

where T_{01} and T_{02} are the stagnation temperatures at inlet and outlet, respectively. For given inlet conditions, the outlet temperature can be determined from the specific enthalpy h_{02}, as follows:

$$\left(W_1 + W_{fuel}\right)h_{02} = W_1 h_{01} + \eta_C W_f h_f \tag{5.65}$$

where W_f is fuel massflow, h_f is the fuel calorific value, and η_C is the combustion efficiency.

The pressure drop across a combustor is nominally 5% at the design point (*dp*). The loss factor F_{PL} can be calculated from Equation 5.47. There are two components, namely the *cold loss* or *aerodynamic loss* (which is constant) and the *hot loss* or *fundamental loss* (which is proportional to $\Delta T/T_{01}$). The pressure loss can be approximated as:

$$\frac{\Delta P}{P_{01}} = \left(F_{cold} + F_{hot}\frac{\Delta T}{T_{01}}\right)Q_1^2 \tag{5.66}$$

where

$$F_{cold} = 0.95\,F_{PL} \qquad F_{hot}[\Delta T/T_{01}]_{dp} = 0.05\,F_{PL}$$

5.4.6 Turbine

Generically, a turbine is a rotor that is driven by air from a high-pressure source as it flows to a lower-pressure volume downstream. Expansion is more efficient than compression and so a turbine would only have one or stages. Performance is defined by a *turbine map* that shows lines for pressure ratio P_{02}/P_{01} versus inlet flow function Q_1 over a range of aerodynamic speed N_{aero} (defined in Equation 5.53). An example is given in Figure 5.4. A separate graph shows pressure ratio versus efficiency. Note also that this information is often plotted as Q_1 versus P_{01}/P_{02} (where the axes are switched).

For polytropic efficiency η_p, the temperature ratio T_{02}/T_{01} is calculated using Equation 5.42:

$$\frac{T_{02}}{T_{01}} = \left(\frac{P_{02}}{P_{01}}\right)^{\frac{n-1}{n}} \tag{5.67}$$

where the polytropic compression index n is defined by:

$$\frac{n-1}{n} = \eta_p\left(\frac{\gamma-1}{\gamma}\right) \tag{5.68}$$

For isentropic efficiency η_i, the temperature ratio is:

$$\frac{T_{02}}{T_{01}} = 1 + \eta_i\left(\left(\frac{P_{02}}{P_{01}}\right)^{\frac{\gamma-1}{\gamma}} - 1\right) \tag{5.69}$$

The fact that the efficiency is less than one means that the gas to cool down is less than would be expected for isentropic flow. Note that polytropic efficiency *is less than* isentropic efficiency for expansion.

The power released by a turbine is equal to the rate at which thermal energy is subtracted from the massflow. The change in temperature can be converted into a change in specific

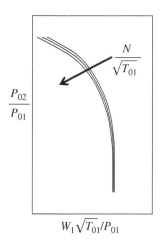
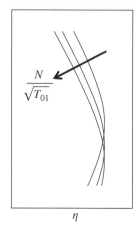

Figure 5.4 Example Turbine Map.

enthalpy [cf. Section 5.6]. Therefore, for an inlet (and outlet) massflow W_1 that is cooled from T_{01} to T_{02}, the turbine power (otherwise known as 'turbine work') is calculated as:

$$J_{turbine} = W_1(h_{01} - h_{02}) \tag{5.70}$$

5.4.7 Nozzle

Nozzle conditions determine the overall output of the engine. For a convergent nozzle (assuming isentropic expansion), the Mach number of the fully expanded jet is:

$$M_J = \sqrt{\frac{2}{\gamma - 1}\left(\left(\frac{P_{01}}{P_\infty}\right)^{\frac{\gamma-1}{\gamma}} - 1\right)} \tag{5.71}$$

where P_{01} is the stagnation pressure at the nozzle inlet and P_∞ is the ambient static pressure at the outlet. The nozzle is choked when the nozzle pressure ratio exceeds the *critical pressure ratio* (CPR):

$$CPR = \left(\frac{\gamma + 1}{2}\right)^{\frac{\gamma}{\gamma-1}} \tag{5.72}$$

In this condition, $M_j = 1$ and the static pressure inside the nozzle is given by:

$$P_1 = \frac{P_{01}}{CPR} \tag{5.73}$$

Otherwise, the static pressure inside the nozzle is equal to the ambient pressure:

$$P_1 = P_\infty \tag{5.74}$$

The corresponding static temperature T_1 can be calculated using Equation 5.39.

Ideally, the gross thrust delivered at the nozzle exit is the sum of pressure and momentum forces:

$$X_G = A(P_1 - P_\infty) + W\nabla \tag{5.75}$$

It is noted that there is no pressure force when the nozzle is unchoked. Using Equations 5.30, 5.31, and 5.38, the flow velocity ∇ is determined as:

$$\nabla = M\sqrt{\gamma RT} = M\sqrt{\gamma RT_0}\sqrt{\frac{T}{T_0}}$$

$$\frac{\nabla}{\sqrt{T_0}} = M\sqrt{\gamma R}\left(1 + \frac{\gamma - 1}{2}M^2\right)^{-\frac{1}{2}} \tag{5.76}$$

Using Equations 5.7, 5.27, and 5.38, the massflow $W = \dot{m}$ is determined as:

$$W_1 = \frac{P}{RT}A\nabla = M\sqrt{\frac{\gamma}{R}}\left(\frac{AP_{01}}{\sqrt{T_{01}}}\right)\left(\frac{P_{01}}{P_1}\right)^{-1}\left(\frac{T_{01}}{T_1}\right)^{\frac{1}{2}}$$

$$\frac{Q_1}{A} = M \sqrt{\frac{\gamma}{R}} \left(1 + \frac{\gamma - 1}{2} M^2\right)^{-\frac{1}{2}\left(\frac{\gamma + 1}{\gamma - 1}\right)} \tag{5.77}$$

where A is the nozzle area and where the flow function Q_1 is defined by Equation 5.48.

5.5 Engine Dynamics

5.5.1 Shaft Speed Variation

Shaft rotation is determined by Newton's second law. The nett power transmission along the shaft is the difference between the power input from the turbine and the power output to the compressor. The constitutive equation for rotation is:

$$\dot{N} = \frac{1}{I} \left(\frac{J_{turbine} - J_{compressor}}{N}\right) \tag{5.78}$$

where I = turbomachine inertia and N = rotational speed.

5.5.2 Massflow Variation

Gas momentum μ is also determined by Newton's second law, based on the static pressure difference ΔP across the control volume and the cross-sectional area A of the control volume:

$$A\Delta P = \dot{\mu} \tag{5.79}$$

Mass is defined by $m = \rho V$, for gas density ρ and fixed volume $V = AL$ (area multiplied by length). Massflow is defined as $W = \dot{m} = \rho A \nabla$, where ∇ is the flow velocity. The derivation is as follows:

$$\mu = m\nabla = m\frac{W}{\rho A} = \frac{V}{A} W = LW$$

$$L\dot{W} = A\Delta P$$

Controversially, ΔP can be construed as the difference between stagnation pressures. The numerical difference between static and stagnation conditions is not huge and the rates of change tend to be relatively similar. In addition, the use of control volume equations in general (operating on homogeneous blocks of stationary air) is not exact and, as such, correction factors will always be applied in order to match the predicted transient response with that observed in a real engine. So, this is a justifiable approximation, as well as being convenient.

Therefore, the rate of change of massflow is here defined as:

$$\dot{W} = K_W (P_{01} - P_{02}) \tag{5.80}$$

where inlet and outlet conditions are labelled as '1' and '2' and where the rate factor is nominally defined as $K_W = A/L$ (area divided by length).

5.5.3 Pressure Variation at Constant Temperature

The Ideal Gas Law gives the relationship between pressure and temperature for fixed volume:

$$PV = mRT \tag{5.81}$$

By differentiation:

$$\dot{P} = \frac{R}{V}\left(m\dot{T} + T\dot{m}\right)$$

$$\dot{P} = \frac{P}{T}\dot{T} + \frac{RT}{V}\dot{m} \tag{5.82}$$

where $\dot{m}$ represents the nett massflow into the control volume, such that $\dot{m} = W_1 - W_2$. The numerical values of P and T represent the mean pressure and mean temperature, respectively, inside the control volume.

As in Section 5.4.2, this can be recast in terms of stagnation conditions. Thus, pressure variation at constant temperature is here defined by:

$$\dot{P}_{02} = K_P(W_1 - W_2) \tag{5.83}$$

where inlet and outlet conditions are labelled as '1' and '2' and where the rate factor is nominally defined as $K_P = RT_0/V$.

5.5.4 Pressure and Temperature Variation

In the general case, coupled variations of pressure and temperature is given by Equation 5.83:

$$\dot{P} = \frac{P}{T}\dot{T} + \frac{RT}{V}\dot{m}$$

The temperature variation is governed by the rate at which fuel is injected and undergoes combustion. This will occur at a very much slower rate than pressure variation, thereby enabling $\dot{P}$ to be calculated as if $\dot{T}$ were zero. So, consider the the energy contained within a fixed volume of gas, which is given by $U = mC_VT$, where T is slowly varying. Its rate of change is approximated as:

$$\dot{U} = \dot{m}C_VT$$

The stand-alone calculation of $\dot{P}$ proceeds, as follows:

$$\dot{P} = \frac{RT}{V}\dot{m} = \left(\frac{R}{V}\right)\frac{\dot{U}}{C_V}$$

Using Mayer's Law ($R = C_P - C_V$) and the Adiabatic Gas Ratio ($\gamma = C_P/C_V$), this becomes:

$$\dot{P} = \frac{\gamma - 1}{V}\dot{U} \tag{5.84}$$

The subsequent calculation of $\dot{T}$ is as follows:

$$\dot{T} = \left(\frac{T}{P}\right)\left[\dot{P} - \frac{RT}{V}\dot{m}\right] \tag{5.85}$$

Again, as in Section 5.4.2, this can be recast in terms of stagnation conditions. Thus, pressure and temperature variations are here defined by:

$$\dot{P} = K_E(W_1 h_{01} - W_2 h_{02}) \tag{5.86}$$

$$\dot{T} = \left(\frac{T_0}{P_0}\right)\left[\dot{P} - K_P(W_1 - W_2)\right] \tag{5.87}$$

where inlet and outlet conditions are labelled as '1' and '2' and where the rate factors are nominally defined as $K_P = RT_0/V$ and $K_E = (\gamma - 1)/V$. Note that the nett flow of mass into the control volume is $W_1 - W_2$ and the nett flow of energy is $W_1 h_{01} - W_2 h_{02}$.

5.6 Engine Models

Engine models can now be assembled, having defined a set of components in Section 5.3 and dynamic equation in Section 5.4. The modelling method will be developed for turbojets and turbofans. For practicality, the level of detail is limited so that a concise explanation can be offered without be drawn into too much detail. The objective is to lay out the fundamental principles of engine composition and component interaction. This should enable the simulation of engine concepts and enable subsequent enhancements by anyone who has interest in specific engine types.

5.6.1 Turbojet Engine

A turbojet contains a single turbomachine (which is a compressor/turbine/shaft combination), combustor and a nozzle, as shown in Figure 5.5. Engine stations are numbered 1 through 6 as the basis for referencing gas properties. Note that it is conventional to draw just half an engine!

Figure 5.5 Turbojet Schematic.

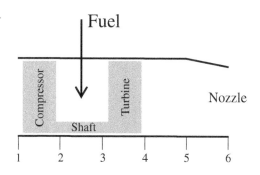

The modelling method is divided into four stages:

- *Specification* imports design parameters and calculated gas conditions in the engine.
- *Initialisation* establishes the initial engine state and the derived parameters for calculation.
- *Physics* calculates gas conditions based on engine states, inlet conditions, and fuel flow.
- *Dynamics* calculates the rates of change of states.

5.6.1.1 Turbojet Specification

Figure 5.6 shows a schematic of the engine. Stations 1–5 hold values of massflow W, stagnation pressure P_0, stagnation temperature T_0, and the associated specific enthalpy h_0, as well as the energy flow (calculated as Wh_0). Engine speed N will be 100%. Inlet massflow W_1 is detemined from the compressor map, for the line that relates pressure ratio to the flow function $Q_1 = W_1\sqrt{T_{01}}/P_{01}$ (as defined in Equation 5.48).

The first pass through the engine model (in Figure 5.7) calculates the massflow, pressure, and temperature at each station, according to a design specification. Typical inlet

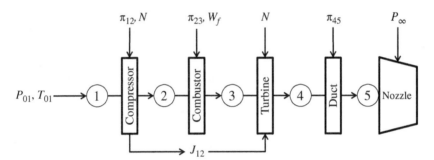

Figure 5.6 Turbojet Specification.

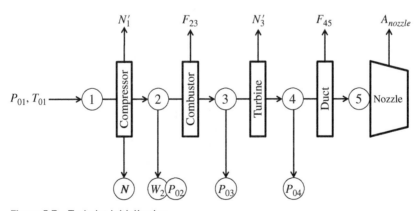

Figure 5.7 Turbojet Initialisation.

conditions are *hot-day sea-level static* (i.e. $P_{01} = 101\,325$ Pa and $T_{01} = 320$ K). Specified pressure ratios are:

$$\pi_{12} = \frac{P_{02}}{P_{01}} \qquad \pi_{23} = \frac{P_{03}}{P_{02}} \qquad \pi_{45} = \frac{P_{05}}{P_{04}} \tag{5.88}$$

The temperature for the compressor is calculated using Equation 5.54:

$$\tau_{12} = \frac{T_{02}}{T_{01}} = \left(\frac{P_{02}}{P_{01}}\right)^{\frac{n-1}{n}} \tag{5.89}$$

The power consumed by the compressor is defined by Equation 5.57:

$$J_{12} = W_1(h_{02} - h_{01}) \tag{5.90}$$

Conversion between temperature and specific enthalpy is defined in Section 5.6. The combustor adds fuel W_f to the airflow, such that:

$$W_3 = W_2 + W_f \tag{5.91}$$

Energy is increased according to Equation 5.65:

$$W_3 h_{03} = W_2 h_{02} + \eta_C W_f h_f \tag{5.92}$$

where h_f is the fuel calorific value (approximately 43 MJ kg^{-1}) and η_C is the combustion efficiency (approximately 0.997). This determines the temperature T_{03} [cf. Section 5.6].

Power to drive the compressor must be generated by the turbine:

$$J_{34} = \frac{J_{12}}{\eta_{mech}} \tag{5.93}$$

where η_{mech} is the mechanical efficiency (nominally 0.99). This matches the change in enthalpy flow across the turbine that is given by Equation 5.70:

$$J_{34} = W_3(h_{03} - h_{04}) \tag{5.94}$$

The temperature ratio across the turbine is now obtained as [cf. Section 5.6]:

$$\tau_{34} = \frac{T_{03}}{T_{04}} \tag{5.95}$$

and the corresponding pressure ratio is derived from Equation 5.67:

$$\pi_{34} = \frac{P_{04}}{P_{03}} = \left(\frac{T_{04}}{T_{03}}\right)^{\frac{n}{n-1}} \tag{5.96}$$

Thus, the propagation of massflow, pressure, and temperature is known. In summary, massflow is W_1 upstream of the combustor and $W_1 + W_f$ downstream. Pressure ratios are defined along the engine and the temperature changes are calculated for compressor, combustor, and turbine. Temperature is constant across the jetpipe. This gives the inlet conditions for the nozzle (as defined in Section 5.4.7).

A summary of engine parameters and gas conditions is given in Tables 5.2 and 5.3, respectively.

Table 5.2 Turbojet Parameters.

Specified Parameters		Derived Parameters	
P_∞	Ambient static pressure	J	Compressor work
P_{01}	Inlet stagnation pressure	τ_{34}	Temperature ratio for turbine
T_{01}	Inlet stagnation temperature	A_{nozzle}	Nozzle area
N	Shaft speed	N_1'	Aerodynamic speed $N/\sqrt{T_{01}}$
π_{12}	Pressure ratio for compressor	N_3'	Aerodynamic speed $N/\sqrt{T_{03}}$
π_{23}	Pressure ratio for combustor	F_{34}	Pressure loss factor for combustor
π_{45}	Pressure ratio for jetpipe	F_{34}	Pressure loss factor for jetpipe

Table 5.3 Derivation of Gas Conditions.

Station	Massflow W	Pressure P_0	Temperature T_0
1	From compressor map	From specification	From specification
2	Equal to Station 1	Derived from π_{12}	Derived from π_{12}
3	Add fuel massflow W_f	Derived from π_{23}	Add energy $W_f h_f$
4	Equal to Station 3	Derived from τ_{34}	τ_{34} from compressor work
5	Equal to Station 4	Derived from π_{45}	Equal to Station 4

5.6.1.2 Turbojet Initialisation

Having obtained the gas conditions at each station, the engine can be initialised for the design specification [cf. Figure 5.5]. inlet conditions are $P_{01} = 101\,325$ Pa and $T_{01} = 320$ K.

Nondimensional engine speed is $N = 1$. The aerodynamic speeds for the compressor N_1' and the turbine N_3' are:

$$N_1' = \frac{N}{\sqrt{T_{01}}} \qquad N_3' = \frac{N}{\sqrt{T_{03}}} \tag{5.97}$$

The pressure loss factors F_{23} and F_{45} are found from Equation 5.47:

$$F_{23} = \frac{1 - \pi_{23}}{Q_2^2} \qquad F_{45} = \frac{1 - \pi_{45}}{Q_4^2} \tag{5.98}$$

where π_{23} and π_{45} are specified pressure ratios.

Nozzle area A_{nozzle} is calculated by Equation 5.77:

$$\frac{Q_5}{A_{nozzle}} = M\sqrt{\frac{\gamma}{R}}\left(1 + \frac{\gamma - 1}{2}M^2\right)^{-\frac{1}{2}\left(\frac{\gamma+1}{\gamma-1}\right)} \tag{5.99}$$

where M is the Mach numebr of the fully expanded jet, R is the gas constant from fuel/air combustion products, and γ is the adiabatic gas ratio [cf. Section 5.6].

The engine states can now be recorded, namely of N, W_2, P_{02}, P_{04}, and T_{03}. In effect, state variables are proxy measures for the energy contained within the engine. Each one is associated with a dynamic process and, as such, it is defined by a differential equation. In other words, these variables provide the basis for calculating how the engine behaves over time.

5.6.1.3 Turbojet Physics

The thermo-physics of the turbojet engine are represented in Figure 5.8. The state variables are injected into the model, together with the inlet conditions P_{01} and T_{01}. State W_2 determines the inlet massflow W_1. The remaining states (N, P_{02}, P_{04}, T_{03}) are all known before calculation starts.

As shown already, the compressor pressure ratio $\pi_{12} = P_{02}/P_{01}$ determines the temperature ratio $\tau_{12} = T_{02}/T_{01}$, thereby providing the gas conditions at Station 2.

For the combustor, the temperature ratio is $\tau_{23} = T_{03}/T_{02}$. The corresponding pressure ratio can be calculated from a modified form of Equation 5.66:

$$\pi_{23} = 1 - (F_{cold} + F_{hot}(\tau_{23} - 1))Q_2^2 \tag{5.100}$$

As before, the addition of fuel is accounted as:

$$W_3 = W_2 + W_f \qquad W_3 h_{03} = W_2 h_{02} + \eta_C W_f h_f \tag{5.101}$$

The turbine compressor pressure ratio $\pi_{34} = P_{04}/P_3$ determines the temperature $\tau_{34} = T_{04}/T_{03}$, thereby providing the gas conditions at Station 4, noting that $W_4 = W_3$.

The jetpipe pressure loss factor is ratio is F_{45}. The actual pressure loss is calculated by Equation 5.47:

$$\pi_{45} = 1 - F_{45}Q_4^2 \tag{5.102}$$

This provides the gas conditions at Station 5, noting that $W_5 = W_4$ and $T_5 = T_4$. These are the inlet conditions for the nozzle (as defined in Section 5.4.7), which enables the calculation of thrust X_G.

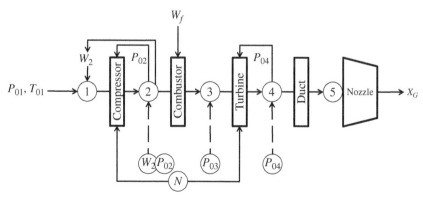

Figure 5.8 Turbojet Physics.

5.6.1.4 Turbojet Dynamics

The last step in the modelling process is to calculate the rates of change of N, W_2, P_{02}, P_{04}, and T_{03}. This follows the approach developed in Section 5.4. For context, the dynamics of the turbojet engine are represented in Figure 5.9.

Variation of shaft speed is given by Equation 5.79:

$$\dot{N} = \frac{1}{I}\left(\frac{J_{34} - J_{12}}{N}\right) \tag{5.103}$$

Variation of mass is given by Equation 5.81:

$$\dot{W} = K_W\left(P'_{02} - P_{03}\right) \tag{5.104}$$

Variation of pressure at constant temperature is given by Equation 5.84:

$$\dot{P}_{04} = K_P\left(W_4 - W'_5\right) \tag{5.105}$$

Combined variation of pressure and temperature is given by Equations 5.87 and 5.88:

$$\dot{P}_{02} = K_E\left(W_2 h_{02} + W_f h_f - W'_3 h_{03}\right) \tag{5.106}$$

$$\dot{T}_{03} = \left(\frac{T_0}{P_0}\right)\left[\dot{P}_{02} - K_P\left(W_2 + W_f - W'_3\right)\right] \tag{5.107}$$

These equations contain modified gas conditions (P'_{02}, W'_3, W'_5) that are calculated in order to create a mismatch with respect to (P_{02}, W_3, W_5) as shaft speed changes. P'_{02} is synthesized from a modified compressor map that plots P_{02}/P_{01} versus $W_1\sqrt{T_{01}}/P_{02}$. A typical map is given in Figure 5.3. A modified map is given in Figure 5.10, showing vertical lines replaced by monotonic curves. The value of W'_3 comes directly from the turbine map and the value of W'_5 comes from Equation 5.78 (that is used to calculate the flow function Q_5 at the nozzle inlet). When the engine is in a steady state, it is clear that $P'_{02} = P_{02}$, $W'_3 = W_3$, and $W'_5 = W_5$.

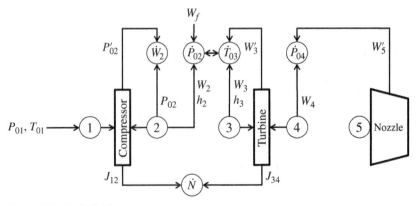

Figure 5.9 Turbojet Dynamics.

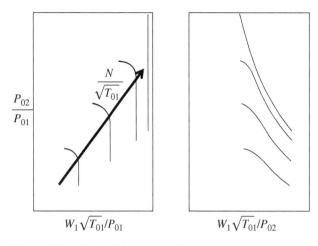

Figure 5.10 Modified Compressor Map.

5.6.2 Turbofan Engine

A turbofan is a two-shaft machine that derives efficiency benefits from being able to optimise the distribution of flow between two separate paths, one through the 'core' (which is a turbojet) and one through the 'bypass', as shown in Figure 5.11.

The modelling philosophy is the same as for the turbojet, except that there are more components. As before, there are four stages:

- *Specification* imports design parameters and calculated gas conditions in the engine.
- *Initialisation* establishes the initial engine state and the derived parameters for calculation.
- *Physics* calculates gas conditions based on engine states, inlet conditions, and fuel flow.
- *Dynamics* calculates the rates of change of states.

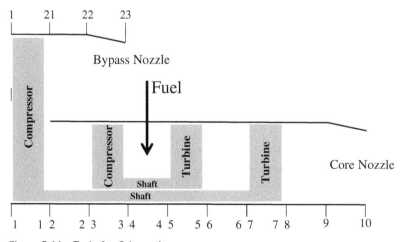

Figure 5.11 Turbofan Schematic.

Rather than repeating the discussion line-by-line from Section 5.5.1, the focus here will be placed on the main points of difference. The design specifies Low Pressure (LP) and High Pressure (HP) systems. The HP system is a turbojet (i.e. a single-shaft turbomachine and a combustor). This is enclosed by a second turbomachine, which incorporates a split compressor with flow separation between core and bypass, as defined by the bypass ratio W_{21}/W_2. The core flow runs through all stages of compression and expansion and then out through the core nozzle. The bypass flow runs out through the bypass nozzle. All component calculations are the same as for the turbojet but the numbering of engine stations has grown to accommodate more components.

Figure 5.12 shows the turbofan specification and Figure 5.13 shows the generation of the derived parameters and the engine states: N_L, N_H, W_2, W_{21}, P_{02}, P_{03}, P_{05}, P_{06}, and T_{04}. A complete summary of engine parameters is given in Table 5.4.

Turbofan physics [Figure 5.14] and turbofan dynamics [Figure 5.15] operate in exactly the same way as for the turbojet and use the same equations, modified to incorporate the new numbering of engine stations. The approach is modular and adaptable, so that alternative architectures can be modelled.

There is much more that could be discussed and a lot more detail that could be developed but the purpose of this section is to present the principles for model-building. The structure of calculation that is implied by the preceding figures should be straightforward to understand but the association of constitutive equations with these block diagrams does require patience and perseverance. However, once this has been done and implemented in software, the principles become obvious!

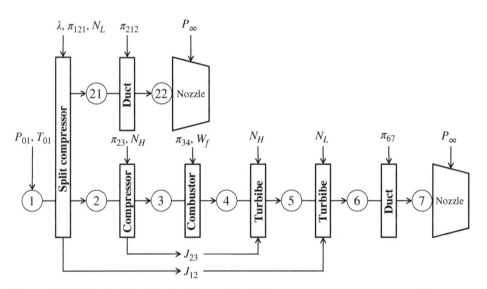

Figure 5.12 Turbofan Specification (Unmixed Cycle).

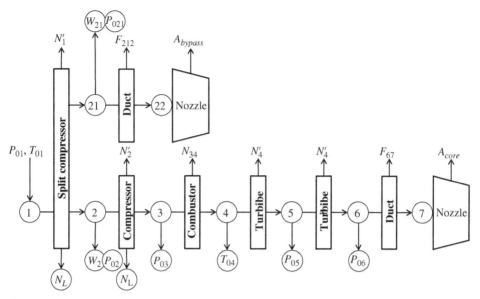

Figure 5.13 Turbofan Initialisation (Unmixed Cycle).

Table 5.4 Turbofan Parameters (Unmixed Cycle).

	Specified Parameters		Derived Parameters
P_∞	Ambient Static Pressure	J_{12}	LP Compressor Work
P_{01}	Inlet stagnation pressure	J_{23}	HP compressor work
T_{01}	Inlet stagnation temperature	τ_{45}	Temperature ratio for HP turbine
N_L	LP shaft speed	τ_{56}	Temperature ratio for LP turbine
N_H	HP shaft speed	A_{bypass}	Bypass nozzle area
λ	Bypass ratio W_{21}/W_2	A_{core}	Core nozzle area
		N'_1	Aerodynamic speed $N_L/\sqrt{T_{01}}$
π_{12}	Pressure ratio for LP inner compressor	N'_2	Aerodynamic speed $N_H/\sqrt{T_{03}}$
π_{121}	Pressure ratio for LP outer compressor	N'_4	Aerodynamic speed $N_H/\sqrt{T_{04}}$
π_{23}	Pressure ratio for HP compressor	N'_5	Aerodynamic speed $N_L/\sqrt{T_{05}}$
π_{34}	Pressure ratio for combustor	F_{34}	Pressure loss factor for combustor
π_{67}	Pressure ratio for core exit	F_{34}	Pressure loss factor for core exit
π_{212}	Pressure ratio for bypass exit	F_{67}	Pressure loss factor for bypass exit

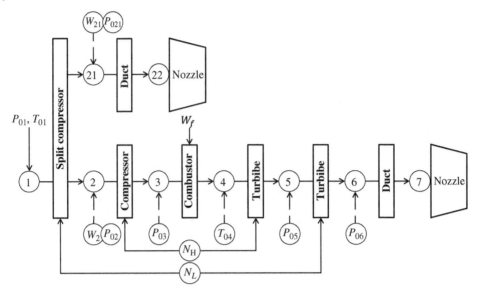

Figure 5.14 Turbofan Physics (Unmixed Cycle).

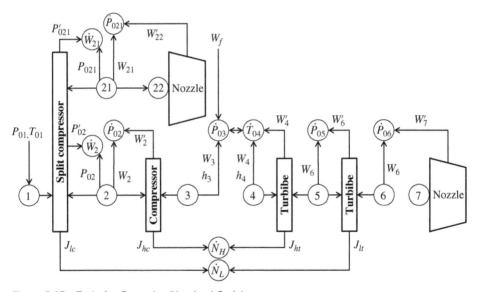

Figure 5.15 Turbofan Dynamics (Unmixed Cycle).

5.7 Gas Properties Data

There have been numerous references to gas properties throughout this chapter, not least the conversion between temperature and specific enthalpy. This final section presents a comprehensive of relevant information and calculations.

5.7.1 Summary of Gas Properties

The equation of state for an ideal gas is expressed by the ideal gas law:

$$PV = mRT \tag{5.108}$$

for a given mass m and gas constant $R = R_0/\mu$, where μ is the mean molecular mass and R_0 is the universal gas constant. The specific heat capacity of the gas is denoted by C_V at *constant volume* and C_P at *constant pressure*. They are inter-related by Mayer's Law and by the adiabatic gas ratio:

$$R = C_P - C_V \qquad \gamma = \frac{C_P}{C_V} \tag{5.109}$$

The specific enthalpy of a gas at a given temperature is the total energy stored per unit mass. For a temperature change from T_1 to T_2, the change in specific enthalpy at constant pressure is often approximated:

$$\Delta h = C_P(T_1)\Delta T \tag{5.110}$$

where $\Delta T = T_2 - T_1$. The specific enthalpy of the gas is actually the integral of specific heat capacity:

$$h(T) = \int_0^T C_P(T)dT \tag{5.111}$$

However, as C_P is difficult to measure at low temperature, it is usual to integrate up from some minimum temperature T_{min} and to assume that C_P is constant below that temperature. So, this equation becomes:

$$h'(T) = C_P(T_{min})T_{min} + \int_{T_{min}}^T C_P(T)dT \tag{5.112}$$

Thus, Equation 5.110 is rewritten as:

$$\Delta h = h'(T_2) - h'(T_1) \tag{5.113}$$

It is the change in specific enthalpy that is important, not the absolute values before and after the change.

5.7.2 Gas Mixtures defined by Mass Fractions

Consider a gas mixture with N components with masses m_i and specific heat capacities C_{Pi} (defined for $i = 1, ..., N$). The corresponding total values m and C_P are given by:

$$m = \sum_{i=1}^N m_i \qquad mC_P = \sum_{i=1}^N m_i C_{Pi} \tag{5.114}$$

Using mass fractions, this becomes:

$$C_P = \sum_{i=1}^N \frac{m_i}{m} C_{Pi} \tag{5.115}$$

For two components, it is usual to define a mass ratio $\lambda = m_2/m_1$, such that:

$$\frac{m_1}{m} = \frac{m_1}{m_1 + m_2} = \frac{1}{1 + \lambda} \qquad \frac{m_2}{m} = \frac{m_2}{m_1 + m_2} = \frac{\lambda}{1 + \lambda} \qquad (5.116)$$

Thus, in this case, Equation 5.114 can be re-expressed as:

$$C_P = \frac{C_{P1} + \lambda C_{P2}}{1 + \lambda} \qquad (5.117)$$

5.7.3 Gas Mixtures defined by Mole Fractions

Mass can be expressed as $m = n\mu$, where n = number of moles of gas and μ = mean molecular mass. Equation 5.114 now can be rewritten in terms of mole fractions (n_i/n):

$$\mu = \sum_{i=1}^{N} \left(\frac{n_i}{n}\right)\mu_i \qquad \mu C_P = \sum_{i=1}^{N} \left(\frac{n_i}{n}\right)\mu_i C_{Pi} \qquad (5.118)$$

Recall that the gas constant is $R = R_0/\mu$, where μ is the mean molecular mass and R_0 is the universal gas constant. So, this equation can be rewritten in terms of specific gas constants:

$$\mu = \sum_{i=1}^{N} \left(\frac{n_i}{n}\right)\mu_i \qquad \frac{C_P}{R} = \sum_{i=1}^{N} \left(\frac{n_i}{n}\right)\frac{C_{Pi}}{R_i} \qquad (5.119)$$

For two components, with a mass ratio λ, the corresponding mole ratio ν is:

$$\nu = \frac{n_2}{n_1} = \left(\frac{m_2}{m_1}\right)\frac{\mu_1}{\mu_2} = \lambda\frac{\mu_1}{\mu_2} \qquad (5.120)$$

Mole fractions are derived as follows:

$$\frac{n_1}{n} = \frac{n_1}{n_1 + n_2} = \frac{1}{1 + \nu} \qquad \frac{n_2}{n} = \frac{n_2}{n_1 + n_2} = \frac{\nu}{1 + \nu} \qquad (5.121)$$

Using Equation 5.118, the mean molecular mass and specific heat capacity are obtained as:

$$\mu = \left(\frac{1}{1+\nu}\right)\mu_1 + \left(\frac{\nu}{1+\nu}\right)\mu_2 \qquad \frac{C_P}{R} = \left(\frac{1}{1+\nu}\right)\frac{C_{P1}}{R_1} + \left(\frac{\nu}{1+\nu}\right)\frac{C_{P2}}{R_2} \qquad (5.122)$$

5.7.4 Dry Air

The chemical composition of dry air is approximated by:

$$\text{Air} = 0.78083N_2 + 0.209460_2 + 000933Ar + 0.00035CO_2 + 0.00002Ne$$

The mean molecular mass of dry air is calculated (from Equation 5.119) with five gas components in Table 5.5 using molecular mass data from NASA-TM-4513. Also, the universal

Table 5.5 Approximate Composition of Dry Air.

Gas	Molecular Mass	Mole Fraction	Mass Component
Nitrogen: N_2	28.01340	0.78083	21.87370
Oxygen: O_2	31.99880	0.20946	6.70247
Argon: Ar	39.94800	0.00933	0.37271
Carbon Dioxide: CO_2	44.00980	0.00035	0.01540
Neon: Ne	20.17970	0.00002	0.00040
	Mean molecular mass (dry air)		**28.96468**

gas constant is quoted as $R_0 = 8314.510$ J. $kmol^{-1}$. K^{-1}. Thus, the gas constant for dry air is calculated as:

$$R = \frac{R_0}{\mu} = 287.05686 \text{ J.kg}^{-1}.\text{K}^{-1} \tag{5.123}$$

where the mean molecular mass is obtained from Table 5.2 [with units kg. $kmol^{-1}$].

5.7.5 Fuel/Air Combustion Products

Most gas turbines run on kerosene with a composition which approximates to $(CH_2)_n$. Combustion chemistry can be idealised to the following reaction:

$$2CH_2 + 3O_2 \longrightarrow 2CO_2 + 2H_2O$$

This ignores unwanted by-products such as carbon monoxide and nitrous oxides that result from dissociation. Pelton (1976) proposed a boundary criterion to determine whether dissociation needs to be included in simulations of combustion chemistry. This boundary is shown in Figure 5.16 and the criterion is stated as a relationship between temperature T measured in Kelvin and pressure P measured in bars [1 bar = 100 000 Pa]:

$$T < 1667 + 101 \log P \tag{5.124}$$

Accordingly, dissociation is neglected for any combination of P and T to the left of the boundary.

In this case, define a fuel/air ratio (FAR), as follows:

$$\lambda = \frac{m_F}{m_A} \tag{5.125}$$

Equation 5.120 gives the corresponding mole ratio:

$$\nu = \frac{n_F}{n_A} = \lambda \frac{\mu_A}{\mu_F} \tag{5.126}$$

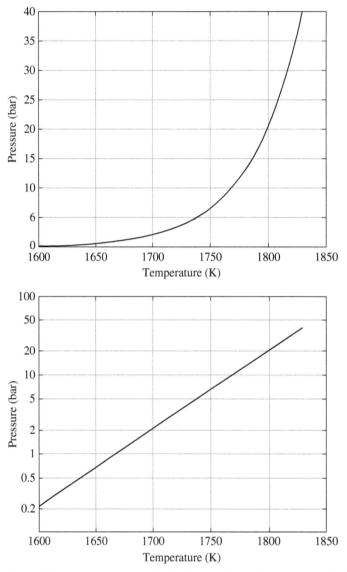

Figure 5.16 Boundary for Negligible Dissociation in Combustion (to Left of the Curve).

Then, Equation 5.17 gives the corresponding mole fractions:

$$\frac{n_A}{n} = \frac{1}{1+\nu} \qquad \frac{n_F}{n} = \frac{\nu}{1+\nu} \tag{5.127}$$

Mean molecular mass and specific heat capacity follow from Equation 5.37:

$$\mu = \left(\frac{n_A}{n}\right)\mu_A + \left(\frac{n_F}{n}\right)\mu_F \qquad \frac{C_P}{R} = \left(\frac{n_A}{n}\right)\frac{C_{PA}}{R_A} + \left(\frac{n_F}{n}\right)\frac{C_{PF}}{F} \tag{5.128}$$

where $\mu_A = 28.96468$ kg. kmol^{-1} for dry air and $\mu_F = 167.31462$ kg. kmol^{-1} for aviation fuel (from data for JET-A in NASA-TM-4513). Finally, the specific gas constant is defined as $R = R_0/\mu$. The adiabatic gas ratio is $\gamma = C_P/C_V$.

Engineering data (believed to have originated somewhere in Rolls-Royce in the early 1960s) is shown in Figure 5.17, with values of C_P over a temperature range from 200 to 2200 K and for values of FAR between 0 and 0.1. Variation in specific enthalpy is obtained from Equation 5.112 with $T_{min} = 200$ K.

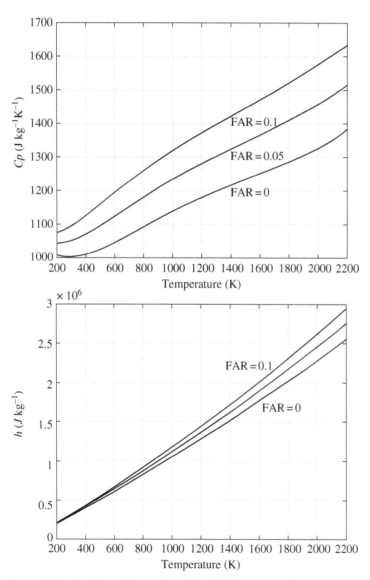

Figure 5.17 Variation of Properties of Fuel/Air Combustion Products.

6

Additional Topics

6.1 Introduction

6.1.1 Expanding the Scope of Volume 2

Chapter 1 introduced a 'simple' flight model and, subsequently, Chapters 2–5 filled in the detail for aerodynamics and propulsion and for equations of motion. There are many other areas of technology that air vehicle modelling could encompass and, by way of introduction, a few of these are included in this chapter. Aircraft structure and aircraft systems represent a substantial proportion of aircraft design effort and, if covered properly, would justify another two textbooks. However, it is useful to summarise the main considerations associated with these areas and to encourage further investigation elsewhere. The additional discussion of mass properties is intended to show the challenge associated with aggregating component properties in order to produce whole-aircraft properties (noting that a large aircraft will comprise tens of thousands of individual components). The specific discussion of fuel mass properties is directly relevant to prediction of flight behaviour under the various operational conditions that the fuel system will find itself, which is important.

6.1.2 What Chapter 6 Includes

Chapter 6 includes:

- Structural Models
- Mass Properties

6.1.3 What Chapter 6 Excludes

Chapter 6 excludes everything else!

6.1.4 Overall Aim

Chapter 6 should provide a sufficient expansion of the previous scope of discussion in Chapters 1–5 so as to make readers aware of some of the larger and more detailed areas of aircraft modelling.

Computational Modelling and Simulation of Aircraft and the Environment: Aircraft Dynamics,
First Edition, Volume II. Dominic J. Diston.
© 2024 John Wiley & Sons Ltd. Published 2024 by John Wiley & Sons Ltd.

6.2 Structural Models

6.2.1 Equations of Motion

The structural dynamics of a flexible vehicle are determined from the inertia, stiffness, and damping of the assembly of structural components (giving a set of modal characteristics), modified by interaction with aerodynamic loads that are generated by geometric distortion of the vehicle structure. The underlying mathematical method is based on coordinate transformations that are used to inter-relate small deflections in translation and/or rotation, together with construction and solution of eigenstructure problems.

The dynamic behaviour of a system is described by a generic equation of the form:

$$M\ddot{q} + D\dot{q} + Eq = Q \tag{6.1}$$

Matrices M, D, and E represent the structural inertia, damping, and stiffness, respectively, vectors q and Q represent generalised displacements and generalised forces, respectively. This equation defines a forced oscillator. Normal modes are established without damping or external forces, such that:

$$M\ddot{q} + Eq = 0 \tag{6.2}$$

Applying Laplace transforms, this becomes:

$$\left(Ms^2 + E\right)q = 0 \tag{6.3}$$

under the assumption of zero displacement at equilibrium. Set $s = j\omega$ in order to obtain the frequency response:

$$\left(-M\omega^2 + E\right)q = 0 \tag{6.4}$$

This compares with the standard eigenstructure problem, where $\omega^2 = 1/\lambda$:

$$(A - \lambda E)q = 0 \tag{6.5}$$

Eigenvalues are determined by solving the characteristic equation:

$$|A - \lambda E| = 0 \tag{6.6}$$

An $N \times N$ matrix has N eigenvalues $[\lambda_n]$, each with an associated eigenvalue $[v_n]$ that is obtained by solving the set of simultaneous linear equations:

$$(A - \lambda_n E)v_n = 0 \tag{6.7}$$

The matrix of eigenvectors is defined by:

$$V = [v_1 \ v_2 \ v_3 \ ... \ v_n] \tag{6.8}$$

This defines an orthogonal transformation between displacements $[q]$ and modal coordinates $[v]$:

$$q = Vv \tag{6.9}$$

If Equation 6.1 is unforced (i.e. $Q = 0$), then:

$$M\ddot{q} = -D\dot{q} - Eq \tag{6.10}$$

This can be manipulated into the following form:

$$\begin{pmatrix} I & 0 \\ 0 & M \end{pmatrix} \begin{pmatrix} \dot{q} \\ \ddot{q} \end{pmatrix} = \begin{pmatrix} 0 & I \\ -E & -D \end{pmatrix} \begin{pmatrix} q \\ \dot{q} \end{pmatrix} \tag{6.11}$$

Applying Laplace transforms, this becomes:

$$\left[s \begin{pmatrix} I & 0 \\ 0 & M \end{pmatrix} - \begin{pmatrix} 0 & I \\ -E & -D \end{pmatrix} \right] \begin{pmatrix} q \\ \dot{q} \end{pmatrix} = 0 \tag{6.12}$$

This can be simplified, as follows:

$$\left(\bar{A} - \lambda\bar{E} \right) \bar{q} = 0 \tag{6.13}$$

where

$$\bar{A} = \begin{pmatrix} I & 0 \\ 0 & M \end{pmatrix} \qquad \bar{E} = \begin{pmatrix} 0 & I \\ -E & -D \end{pmatrix} \qquad \bar{q} = \begin{pmatrix} q \\ \dot{q} \end{pmatrix}$$

6.2.2   Coordinate Transformations

The principle of coordinate transformation can be illustrated for small perturbations, as in Figure 6.1. This shows the zx-plane, with vertical translations z and pitch rotations θ of a rigid element. Pitch angles are measured as right-handed rotations about the y-axis (using standard convention). These can be considered as generalised displacements, which are associated with generalised forces, namely forces F and moments M.

The mathematical transformations between displacements measured at two reference points (labelled as 0 and 1) are, as follows:

$$\begin{pmatrix} z \\ \theta \end{pmatrix}_1 = \begin{pmatrix} 1 & -x \\ 0 & 1 \end{pmatrix} \begin{pmatrix} z \\ \theta \end{pmatrix}_0, \qquad \begin{pmatrix} F \\ M \end{pmatrix}_0 = \begin{pmatrix} 1 & 0 \\ -x & 1 \end{pmatrix} \begin{pmatrix} F \\ M \end{pmatrix}_1, \tag{6.14}$$

In effect, this is lever-arm compensation for a given offset x.

Equivalently, these relationships can be written as follows:

$$q_1 = T_{10}q_0 \qquad Q_0 = T_{10}^T Q_1 \tag{6.15}$$

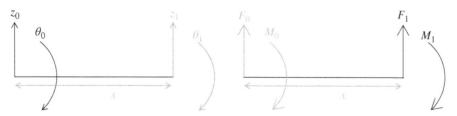

Figure 6.1   Simple Transformation of Displacements and Forces.

where q is a vector of generalised displacements, Q is a vector of generalised forces, and T_{10} is a transformation matrix:

$$T_{10} = \begin{pmatrix} 1 & -x \\ 0 & 1 \end{pmatrix}$$

The subscripts indicate the reference point at which displacements and forces are measured. This is justified by the calculation of work done against a force:

$$W = Q_1^T q_1 = Q_1^T (T_{10} q_0) = (Q_1^T T_{10}) q_0 = Q_0^T q_0 \tag{6.16}$$

This scheme can be extended to three dimensions, where point 1 is located by a position vector r_{10} relative to point 0 and the associated frame xyz_0:

$$r_{10} = \begin{pmatrix} x_{10} \\ y_{10} \\ z_{10} \end{pmatrix} \tag{6.17}$$

Displacements are defined by translation and rotation:

$$\delta = \begin{pmatrix} x \\ y \\ z \end{pmatrix} \qquad \sigma = \begin{pmatrix} \varphi \\ \theta \\ \psi \end{pmatrix} \tag{6.18}$$

Subscripts are used in order to designate the appropriate reference points. Revised transformations are as follows:

$$\delta_1 = \delta_0 - r_{10} \times \sigma_0 \qquad \sigma_1 = \sigma_0 \tag{6.19}$$

Equivalently:

$$\delta_1 = \delta_0 - \Omega_{10}\sigma_0 \qquad \sigma_1 = \sigma_0 \tag{6.20}$$

where

$$\Omega_{10} = \begin{pmatrix} 0 & z_{10} & -y_{10} \\ -z_{10} & 0 & x_{10} \\ y_{10} & -x_{10} & 0 \end{pmatrix} \tag{6.21}$$

Using generalised coordinates, a single transformation can be defined:

$$q_1 = T_{10} q_0 \tag{6.22}$$

where

$$q_1 = \begin{pmatrix} \delta_1 \\ \sigma_1 \end{pmatrix} \qquad q_0 = \begin{pmatrix} \delta_0 \\ \sigma_0 \end{pmatrix} \qquad T_{10} = \begin{pmatrix} I & \Omega_{10} \\ 0 & I \end{pmatrix}$$

6.2.3 Coupled Structure

Consider the example of a coupled structure shown in Figure 6.2. It has three components (each with mass m and inertia J) and it is free to move in pitch (θ) and heave (z). The compact central component is the reference body with generalised coordinates (z_0, θ_0) measured with respect to an external datum. The left-side and right-side linear components are attached via elastic couplings that allow deflections in pitch, $\delta\theta_1$ and $\delta\theta_2$, respectively. The length of each side component is defined as l.

Thus, for small deflections, the required coordinate transformations are:

$$\begin{pmatrix} z_1 \\ \theta_1 \end{pmatrix} = \begin{pmatrix} 1 & l & l \\ 0 & 1 & 1 \end{pmatrix} \begin{pmatrix} z_0 \\ \theta_0 \\ \Delta\theta_1 \end{pmatrix} \qquad \begin{pmatrix} z_2 \\ \theta_2 \end{pmatrix} = \begin{pmatrix} 1 & -l & -l \\ 0 & 1 & 1 \end{pmatrix} \begin{pmatrix} z_0 \\ \theta_0 \\ \Delta\theta_2 \end{pmatrix}$$

$$(6.23)$$

The kinetic energy of this system is calculated as:

$$T = \frac{1}{2} \begin{pmatrix} z_0 \\ \theta_0 \end{pmatrix}^T \begin{pmatrix} m_0 & 0 \\ 0 & J_0 \end{pmatrix} \begin{pmatrix} z_0 \\ \theta_0 \end{pmatrix} + \frac{1}{2} \begin{pmatrix} z_1 \\ \theta_1 \end{pmatrix}^T \begin{pmatrix} m_1 & 0 \\ 0 & J_1 \end{pmatrix} \begin{pmatrix} z_1 \\ \theta_1 \end{pmatrix} + \frac{1}{2} \begin{pmatrix} z_2 \\ \theta_2 \end{pmatrix}^T \begin{pmatrix} m_2 & 0 \\ 0 & J_2 \end{pmatrix} \begin{pmatrix} z_2 \\ \theta_2 \end{pmatrix}$$

$$(6.24)$$

After some algebraic manipulation, this becomes:

$$T = \frac{1}{2} \begin{pmatrix} z_0 \\ \theta_0 \\ \Delta\theta_1 \end{pmatrix}^T \begin{pmatrix} m_0 + m_1 + m_2 & (m_1 - m_2)l & m_1 l & -m_2 l \\ (m_1 - m_2)l & J_0 + J_1 + J_2 + (m_1 + m_2)l^2 & J_1 + m_1 l^2 & J_2 + m_2 l^2 \\ m_1 l & J_1 + m_1 l^2 & J_1 + m_1 l^2 & 0 \\ -m_2 l & J_2 + m_2 l^2 & 0 & J_2 + m_2 l^2 \end{pmatrix} \begin{pmatrix} z_0 \\ \theta_0 \\ \Delta\theta_1 \end{pmatrix}$$

... and this is just for a simple structure comprising three simple components! Accordingly, in this particular subject area, matrix algebra is invariably treated as 'matrix algebra', without opting for element-by-element expansions.

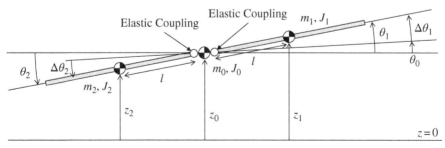

Figure 6.2 Three-Component Coupled Structure.

A rather more streamlined approach is to state the generalised inertia matrices for the structural components, together with the generalised coordinates that define their motion as follows:

$$A_0 = \begin{pmatrix} m_0 & 0 \\ 0 & J_0 \end{pmatrix} \qquad A_1 = \begin{pmatrix} m_1 & 0 \\ 0 & J_1 \end{pmatrix} \qquad A_2 = \begin{pmatrix} m_2 & 0 \\ 0 & J_2 \end{pmatrix} \qquad (6.25)$$

$$\boldsymbol{q}_0 = \begin{pmatrix} z_0 \\ \theta_0 \end{pmatrix} \qquad \boldsymbol{q}_1 = \begin{pmatrix} z_1 \\ \theta_1 \end{pmatrix} \qquad \boldsymbol{q}_2 = \begin{pmatrix} z_2 \\ \theta_2 \end{pmatrix} \qquad (6.26)$$

The coordinate transformations can be written as:

$$\boldsymbol{q}_1 = T_{10}\boldsymbol{q}_0 + S_1 \Delta q_1 \qquad \boldsymbol{q}_2 = T_{20}\boldsymbol{q}_0 + S_2 \Delta q_2 \qquad (6.27)$$

where

$$T_{10} = \begin{pmatrix} 1 & l \\ 0 & 1 \end{pmatrix} \qquad S_1 = \begin{pmatrix} l \\ 1 \end{pmatrix} \qquad T_{20} = \begin{pmatrix} 1 & -l \\ 0 & 1 \end{pmatrix} \qquad S_2 = \begin{pmatrix} -l \\ 1 \end{pmatrix}$$

This leads to the following consolidation:

$$T = \frac{1}{2}(\boldsymbol{q}_0 \ \boldsymbol{q}_1 \ \boldsymbol{q}_2) \begin{pmatrix} A_0 & 0 & 0 \\ 0 & A_1 & 0 \\ 0 & 0 & A_2 \end{pmatrix} \begin{pmatrix} \boldsymbol{q}_0 \\ \boldsymbol{q}_1 \\ \boldsymbol{q}_2 \end{pmatrix} \qquad (6.28)$$

After transforming coordinates, this can be recast in a 'better' computational form:

$$T = \frac{1}{2}(\boldsymbol{q}_0 \ \delta q_1 \ \delta q_2) \begin{pmatrix} A_0 + T_{10}^T A_1 T_{10} + T_{20}^T A_2 T_{20} & T_{10}^T A_1 S_1 & T_{20}^T A_2 S_2 \\ S_1^T A_1 T_{10} & S_1^T A_1 S_1 & 0 \\ S_2^T A_2 T_{20} & 0 & S_2^T A_2 S_2 \end{pmatrix} \begin{pmatrix} \boldsymbol{q}_0 \\ \delta q_1 \\ \delta q_2 \end{pmatrix}$$
$$(6.29)$$

Herein lies the essence of airframe modelling and it provides the basis for describing wing-fuselage structures ... which is the next iteration.

6.2.4 Wing-Fuselage Structure

In order to establish the transformations for a multi-body structure, consider the wing-fuselage configuration shown in Figure 6.3. This comprises five rigid components, labelled 0–4. Each has its own *xyz*-frame located at a datum point, thereby defining position and orientation. For convenience, the fuselage datum is designated as the datum for the whole system and, in this context, it is called the *rigid body*. Each fuselage-wing interface is located at the relevant wing datum and is treated as an articulated joint about which the wing can rotate in two degrees of freedom (to allow bending and torsion). Each wing incorporates a plain flap that can rotate in one degree of freedom (about its hinge line).

As an extension of the basic method of coordinate transformations (given in the previous section), it is necessary to account for the local orientation of each component. Also, in

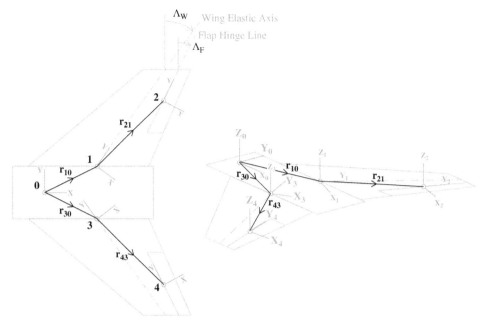

Figure 6.3 Simple Model of Wing-Fuselage Structure.

addition to displacements of each component datum, elastic deflections will be included. These will be modelled as incremental rotations in each of the elastic degrees of freedom that are defined for the particular component. Note that more accurate structural models could have tens, hundreds, or thousands of degrees of freedom; for planar structures, these would account for deflections orthogonal to the structure.

Following the method discussed already, the basic coordinate transformation for the starboard wing is given by:

$$q_1 = T_{10}q_0 \tag{6.30}$$

Using a generalised form of what was presented previously, this becomes:

$$q_1 = T_{10}q_0 + S_1 \Delta q_1 \tag{6.31}$$

where

$$q_1 = \begin{pmatrix} \delta_1 \\ \sigma_1 \end{pmatrix} \qquad \Delta q_1 = \begin{pmatrix} \Delta\delta_1 \\ \Delta\sigma_1 \end{pmatrix} \qquad q_0 = \begin{pmatrix} \delta_0 \\ \sigma_0 \end{pmatrix} \qquad T_{10} = \begin{pmatrix} R^{10} & R^{10}\Omega_{10} \\ 0 & R^{10} \end{pmatrix}$$

In this formulation, $\Delta\delta$ and $\Delta\sigma$ denote elastic deflections in translation and rotation, respectively. Note that the transformation T_{10} now contains a rotation from xyz_0 and xyz_1, which is denoted by R^{10}. This is repeated for component 2 (which is the starboard aileron):

$$q_2 = T_{21}q_1 + S_2 \Delta q_2 \tag{6.32}$$

The transformations for the fuselage and right wing/aileron assembly are combined, as follows:

$$
\begin{pmatrix} q_0 \\ q_1 \\ q_2 \end{pmatrix} = \begin{pmatrix} I & 0 & 0 \\ T_{10} & S_1 & 0 \\ T_{21}T_{10} & T_{21} & S_2 \end{pmatrix} \begin{pmatrix} q_0 \\ \Delta q_1 \\ \Delta q_2 \end{pmatrix} \tag{6.33}
$$

where

$$
T_{10} = \begin{pmatrix} R^{10} & R^{10}\Omega_{10} \\ 0 & R^{10} \end{pmatrix} \qquad T_{21} = \begin{pmatrix} R^{21} & R^{21}\Omega_{21} \\ 0 & R^{21} \end{pmatrix}
$$

In this example, the relative orientation of components is defined by elementary rotations [as discussed at length in Chapter 2]:

$$
R^{10} = R^Z(\Lambda_W) \qquad R^{10} = R^Z(\Lambda_W - \Lambda_F) \tag{6.34}
$$

The kinetic energy of the structure is based on the combined inertia of all structural components. Referring to the fuselage and starboard wing/aileron assembly, this is calculated as:

$$
KE = \frac{1}{2}\begin{pmatrix} \dot{q}_0 \\ \dot{q}_1 \\ \dot{q}_2 \end{pmatrix}^T \begin{pmatrix} M_0 & 0 & 0 \\ 0 & M_1 & 0 \\ 0 & 0 & M_2 \end{pmatrix} \begin{pmatrix} \dot{q}_0 \\ \dot{q}_1 \\ \dot{q}_2 \end{pmatrix} \tag{6.35}
$$

where M_n denotes the *inertia matrix* of the nth component with respect to its local axis system.

Transforming coordinates, the resulting expression has the form:

$$
KE = \frac{1}{2}\begin{pmatrix} \dot{q}_0 \\ \Delta\dot{q}_1 \\ \Delta\dot{q}_2 \end{pmatrix}^T \begin{pmatrix} M_{00} & M_{01} & M_{02} \\ M_{10} & M_{11} & M_{12} \\ M_{20} & M_{21} & M_{22} \end{pmatrix} \begin{pmatrix} \dot{q}_0 \\ \Delta\dot{q}_1 \\ \Delta\dot{q}_2 \end{pmatrix} \tag{6.36}
$$

According to Equation 6.24, this becomes:

$$
KE = \frac{1}{2}\begin{pmatrix} \dot{q}_0 \\ \Delta\dot{q}_1 \\ \Delta\dot{q}_2 \end{pmatrix}^T \begin{pmatrix} M_0 + T_{01}M_1T_{10} + T_{02}M_2T_{20} & T_{01}M_1S_1 + T_{02}M_2T_{21} & T_{02}M_2S_2 \\ S_1^T M_1T_{10} + T_{12}M_2T_{20} & S_1^T M_1S_1 + T_{12}M_2T_{21} & T_{12}M_2S_2 \\ S_2^T M_2T_{20} & S_2^T M_2T_{21} & S_2^T M_2S_2 \end{pmatrix} \begin{pmatrix} \dot{q}_0 \\ \Delta\dot{q}_1 \\ \Delta\dot{q}_2 \end{pmatrix} \tag{6.37}
$$

For convenience, the following shorthand has been employed:

$$
T_{20} = T_{21}T_{10} \tag{6.38}
$$

This defines the coupled inertia of structural components and, in this context, coordinate transformations are often called *inertia couplings*.

The strain energy for the structure is based on a *stiffness matrix* and is calculated as follows:

$$SE = \frac{1}{2} \begin{pmatrix} q_0 \\ \Delta q_1 \\ \Delta q_2 \end{pmatrix}^T \begin{pmatrix} 0 & 0 & 0 \\ 0 & E_1 & 0 \\ 0 & 0 & E_2 \end{pmatrix} \begin{pmatrix} q_0 \\ \Delta q_1 \\ \Delta q_2 \end{pmatrix} \tag{6.39}$$

where E_n represent the stiffness of the nth component, i.e. the generalised force per unit of generalised displacement.

In this example, which uses rigid components, 'stiffness' applies to the elastic degrees of freedom at the component datum. The stiffness matrix is block diagonal, such that elastic displacements of one component only produce forces in that component. Note that the rigid body is unconstrained and, therefore, it is displaced without generating any force. In other words, it has no stiffness and therefore the appropriate matrix elements are all zero.

Following the same method, the model can be expanded to include all five components of the wing-fuselage structure in Figure 6.3. Thus, Equation 6.33 must be replaced by:

$$\begin{pmatrix} q_0 \\ q_1 \\ q_2 \\ q_3 \\ q_4 \end{pmatrix} = \begin{pmatrix} I & 0 & 0 & 0 & 0 \\ T_{10} & S_1 & 0 & 0 & 0 \\ T_{20} & T_{21} & S_2 & 0 & 0 \\ T_{30} & 0 & 0 & S_3 & 0 \\ T_{40} & 0 & 0 & T_{43} & S_4 \end{pmatrix} \begin{pmatrix} q_0 \\ \Delta q_1 \\ \Delta q_2 \\ \Delta q_3 \\ \Delta q_4 \end{pmatrix} \tag{6.40}$$

Kinetic energy is now calculated in the form:

$$KE = \frac{1}{2} \begin{pmatrix} \dot{q}_0 \\ \Delta \dot{q}_1 \\ \Delta \dot{q}_2 \\ \Delta \dot{q}_3 \\ \Delta \dot{q}_4 \end{pmatrix} \begin{pmatrix} M_{00} & M_{01} & M_{02} & M_{03} & M_{04} \\ M_{10} & M_{11} & M_{12} & 0 & 0 \\ M_{20} & M_{21} & M_{22} & 0 & 0 \\ M_{30} & 0 & 0 & M_{33} & M_{34} \\ M_{40} & 0 & 0 & M_{43} & M_{44} \end{pmatrix} \begin{pmatrix} \dot{q}_0 \\ \Delta \dot{q}_1 \\ \Delta \dot{q}_2 \\ \Delta \dot{q}_3 \\ \Delta \dot{q}_4 \end{pmatrix} \tag{6.41}$$

Individual matrix elements M_{ij} are similar in form to those contained in Equation 6.27 and the zero elements exist because there is no direct coupling between left and right wing structures. The strain energy (Equation 6.39) is now calculated in the form:

$$SE = \frac{1}{2} \begin{pmatrix} q_0 \\ \Delta q_1 \\ \Delta q_2 \\ \Delta q_3 \\ \Delta q_4 \end{pmatrix} \begin{pmatrix} 0 & 0 & 0 & 0 & 0 \\ 0 & E_1 & 0 & 0 & 0 \\ 0 & 0 & E_2 & 0 & 0 \\ 0 & 0 & 0 & E_3 & 0 \\ 0 & 0 & 0 & 0 & E_4 \end{pmatrix} \begin{pmatrix} q_0 \\ \Delta q_1 \\ \Delta q_2 \\ \Delta q_3 \\ \Delta q_4 \end{pmatrix} \tag{6.42}$$

The dynamic behaviour of the multi-body structure is defined by:

$$M\Delta\ddot{q} + D\Delta\dot{q} + E\Delta q = \Delta Q \tag{6.43}$$

Matrices M, D, and E represent the structural inertia, damping, and stiffness, respectively. For convenience and compactness, the generalised forces and displacements are written as:

$$\Delta q = \begin{pmatrix} q_0 \\ \Delta q_1 \\ \Delta q_2 \\ \vdots \\ \Delta q_n \end{pmatrix} \qquad \Delta Q = \begin{pmatrix} Q_0 \\ \Delta Q_1 \\ \Delta Q_2 \\ \vdots \\ \Delta Q_n \end{pmatrix} \tag{6.44}$$

These quantities combine rigid body displacements and elastic displacements of the structure.

Equation 6.34 defines a forced oscillator of the type described by Equation 6.1. Following the method given in Section 6.1, this leads to N eigenvalues λ_n and N eigenvectors v_n that are inter-related as follows:

$$(A - \lambda_n E)v_n = 0 \tag{6.45}$$

The matrix of eigenvectors is defined by:

$$V = \begin{bmatrix} v_1 & v_2 & v_3 & \cdots & v_n \end{bmatrix} \tag{6.46}$$

This defines an orthogonal transformation between displacements Δq and modal coordinates Δv:

$$\Delta q = V\Delta v$$

In addition, because mass and stiffness matrices are symmetric, this transformation has the special property of *diagonalisation*, such that

$$V^T M V = \begin{pmatrix} m_1 & 0 & \cdots & 0 \\ 0 & m_2 & \cdots & 0 \\ \vdots & \vdots & \ddots & \vdots \\ 0 & 0 & \cdots & m_n \end{pmatrix} \qquad V^T E V = \begin{pmatrix} e_1 & 0 & \cdots & 0 \\ 0 & e_2 & \cdots & 0 \\ \vdots & \vdots & \ddots & \vdots \\ 0 & 0 & \cdots & e_n \end{pmatrix} \tag{6.47}$$

Thus, the Nth order system of equations is reduced to N independent 1st order systems, such that $M\Delta\ddot{q} + E\Delta q = 0$ becomes:

$$m_i \Delta\ddot{v}_i + e_i \Delta v_i = 0 \tag{6.48}$$

for $n = 1, 2, 3, ..., N$.

At this point, structural damping can be re-introduced:

$$m_i \Delta\ddot{v}_i + d\Delta\dot{v}_i + e_i \Delta v_i = 0 \tag{6.49}$$

The characteristic equation is obtained by Laplace transform:

$$m_i s^2 + ds + e_i = 0 \tag{6.50}$$

Critical damping corresponds with the limiting condition where the roots of this equation are real:

$$[d_i]_{crit} = 2\sqrt{m_i e_i} \tag{6.51}$$

It is conventional to specify modal damping as a percentage of critical damping. Aerospace structures tend to be lightweight structures with relatively high stiffness and, when not immersed in air, very low damping. While that actual damping would need to be verified by measurement, a reasonable first estimate would be around 1% of critical damping, i.e.

$$d_i \approx 0.01 [d_i]_{crit} \tag{6.52}$$

It is usual to provide an overview of non-zero inertia and stiffness, as in Table 6.1. This provides a short-hand method for interpreting how components are inter-connected.

With this method, it is possible to continue extending the structural model by partitioning the structure and adding new components. This is illustrated in Figure 6.4, where the wings have been split into inboard and outboard section, each with a trailing-edge flap. It is worth noting that this allows the elastic axis to be specified separately for each major component. In Figure 6.3, the wing elastic axis is swept back at a constant angle from root to tip: in Figure 6.4, the elastic axis of the inner wing is unswept. The corresponding structures for inertia and stiffness matrices are presented in Table 6.2.

Table 6.1 Internal Layout of Inertia and Stiffness Matrices.

		0	1	2	3	4	0	1	2	3	4
		RIGID BODY	STBD WING	STBD ALRN	PORT WING	PORT ALRN	RIGID BODY	STBD WING	STBD ALRN	PORT WING	PORT ALRN
0	RIGID BODY	X	X	X	X	X					
1	STBD WING	X	X	X				X			
2	STBD FLAP	X	X	X					X		
3	PORT WING	X			X	X				X	
4	PORT FLAP	X			X	X					X
				Inertia Matrix					**Stiffness Matrix**		

Table 6.2 Extended Inertia and Stiffness.

		0	1	2	3	4	5	6	7	8
		RIGID BODY	STBD IB WING	STBD IB ALRN	STBD OB WING	STBD OB ALRN	PORT IB WING	PORT IB ALRN	PORT OB WING	PORT OB ALRN
0	RIGID BODY	X	X	X	X	X	X	X	X	X
1	STBD IB WING	X	X	X	X					
2	STBD IB ALRN	X	X	X						
3	STBD OB WING	X	X		X	X				
4	STBD OB ALRN	X			X	X				
5	PORT IB WING	X					X	X	X	
6	PORT IB ALRN	X					X	X		
7	PORT OB WING	X					X		X	X
8	PORT OB ALRN	X							X	X

Inertia Matrix

0	RIGID BODY									
1	STBD IB WING		X							
2	STBD IB ALRN			X						
3	STBD OB WING				X					
4	STBD OB ALRN					X				
5	PORT IB WING						X			
6	PORT IB ALRN							X		
7	PORT OB WING								X	
8	PORT OB ALRN									X

Stiffness Matrix

Figure 6.4 Extended Wing-Fuselage Model.

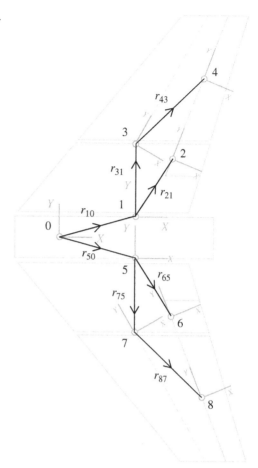

6.2.5 Whole Aircraft Structure

Ultimately, the method of inertia coupling extends to encompass the whole aircraft. The principle remains the same throughout. The root is defined as the 'rigid body' and the wings and fuselage (forward and aft) branch from it. The aft fuselage is the root for the empennage attachments, typically vertical and horizontal tail structures. In some configurations, the horizontal tail is mounted on a nacelle at the top of the vertical tail. In addition, ailerons and flaps are attached to the wings, the rudder is attached to the vertical tail and the elevators are attached to the horizontal tail.

A whole aircraft schematic is presented in Figure 6.5 and the layout of the associated inertia matrix is presented in Table 6.3. The stiffness matrix has a block diagonal layout, as in the case of the wing-fuselage combination. Normal modes can be established by solving the eigenvalue problem and, while avoiding the technical detail in setting up the problem, indicative results are given in Figure 6.6.

Component Indices

0:	Reference Object
1:	F
2:	A
3:	B
4:	W1
5:	F1
6:	W2
7:	F2
8:	W3
9:	F3
10:	W4
11:	F4
12:	V
13:	R
14:	N
15:	H1
16:	E1
17:	H2
18:	E2

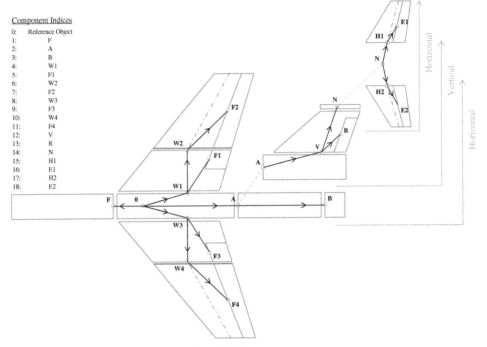

Figure 6.5 Extended Wing-Body-Fuselage Model.

6.3 Mass Distribution

6.3.1 Mass Properties

As introduced in Section 2.2.1, so-called *mass properties* comprise mass (m), moments of inertia (J_x, J_y, J_z), products of inertia (J_{xy}, J_{yz}, J_{zx}) and the coordinates of the centre of gravity ($\bar{x}, \bar{y}, \bar{z}$). All of these quantities derive from summations performed on a mass distribution, i.e. a set of mass elements dm located at coordinates (x, y, z). The centre of gravity is determined from the equations:

$$m = \int dm \qquad m\bar{x} = \int x\,dm \qquad m\bar{y} = \int y\,dm \qquad m\bar{z} = \int z\,dm \qquad (6.53)$$

If coordinates are measured from the CG, then the CG lies at the origin and therefore:

$$\int x\,dm = 0 \qquad \int y\,dm = 0 \qquad \int z\,dm = 0 \qquad (6.54)$$

The kinetic energy associated with each mass element can be written as:

$$dE = \frac{1}{2}V^T dm\,V \qquad (6.55)$$

Table 6.3 Internal Layout of Aircraft Inertia Matrix.

		0	1	2	3	4	5	6	7	8	9	10	11	12	13	14	15	16	17
0	RIGID BODY	X	X	X	X	X	X	X	X	X	X	X	X	X	X	X	X	X	X
F	FRONT FUSELAGE	X	X																
2	AFT FUSELAGE	X		X															
3	STBD IB WING	X			X	X	X												
4	STBD IB ALRN	X			X	X		X											
5	STBD OB WING	X			X		X	X											
6	STBD OB ALRN	X				X	X	X											
7	PORT IB WING	X							X	X	X								
8	PORT IB ALRN	X							X	X		X							
9	PORT OB WING	X							X		X	X							
10	PORT OB ALRN	X								X	X	X							
11	VERT TAIL	X											X	X	X	X	X	X	X
12	RUDDER	X											X	X					
13	NACELLE	X											X		X	X	X	X	X
14	STBD HRZN TAIL	X											X		X	X	X		
15	STBD ELEV	X											X		X	X	X		
16	PORT HRZN TAIL	X											X		X			X	X
17	PORT ELEV	X											X		X			X	X

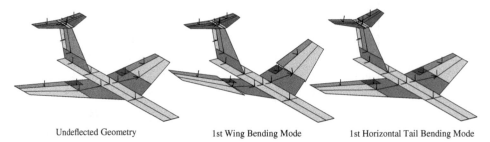

Undeflected Geometry   1st Wing Bending Mode   1st Horizontal Tail Bending Mode

Figure 6.6 Selected Aircraft Normal Modes.

where the velocity V is determined by the angular velocity ω and by the position vector $p = (x, y, z)^T$ of the mass element. Thus,

$$V = p \times \omega = P\omega \qquad (6.56)$$

where the cross-product matrix is:

$$P = \begin{pmatrix} 0 & -z & y \\ z & 0 & -x \\ -y & x & 0 \end{pmatrix} \qquad (6.57)$$

The kinetic energy becomes:

$$dE = \frac{1}{2}\omega^T dJ \omega \qquad (6.58)$$

where the inertia matrix is:

$$dJ = P^T \, dm \, P = P^T P \, dm \qquad (6.59)$$

This is expanded to give:

$$dJ = \begin{pmatrix} 0 & z & -y \\ -z & 0 & x \\ y & -x & 0 \end{pmatrix} \begin{pmatrix} 0 & -z & y \\ z & 0 & -x \\ -y & x & 0 \end{pmatrix} dm = \begin{pmatrix} y^2 + z^2 & -zx & -xy \\ -zx & z^2 + x^2 & -yz \\ -xy & -yz & x^2 + y^2 \end{pmatrix} dm$$

The total inertia matrix are obtained by integration:

$$J = \begin{pmatrix} \int (y^2 + z^2)\,dm & -\int zx\,dm & -\int xy\,dm \\ -\int zx\,dm & \int (z^2 + x^2)\,dm & -\int yz\,dm \\ -\int xy\,dm & -\int yz\,dm & \int (x^2 + y^2)\,dm \end{pmatrix} \qquad (6.60)$$

In general, the position of a mass element can be measured as:

$$p = \bar{p} + \Delta p \qquad or \qquad \begin{pmatrix} x \\ y \\ z \end{pmatrix} = \begin{pmatrix} \bar{x} \\ \bar{y} \\ \bar{z} \end{pmatrix} + \begin{pmatrix} \Delta x \\ \Delta y \\ \Delta z \end{pmatrix} \qquad (6.61)$$

which is the CG position $\bar{p}$ plus the offset position from the CG to the mass element Δp. The inertia matrix [Equation 5.58] is then re-expressed as:

$$dJ = (\bar{P} + \Delta P)^T (\bar{P} + \Delta P)\, dm \qquad (6.62)$$

where the cross-product matrices are:

$$\bar{P} = \begin{pmatrix} 0 & -\bar{z} & \bar{y} \\ \bar{z} & 0 & -\bar{x} \\ -\bar{y} & \bar{x} & 0 \end{pmatrix} \qquad \Delta P = \begin{pmatrix} 0 & -\Delta z & \Delta y \\ \Delta z & 0 & -\Delta x \\ -\Delta y & \Delta x & 0 \end{pmatrix} \qquad (6.63)$$

Thus, the total inertia matrix is determined as:

$$J = \int \left(\bar{P}^T \bar{P} + \bar{P}^T \Delta P + \Delta P^T \bar{P} + \Delta P^T \Delta P \right) dm$$

where $\bar{P}$ is constant and ΔP varies with position (and therefore is integrated). This reduces to:

$$J = \bar{P}^T \bar{P} \int dm + \bar{P}^T \left(\int \Delta P\, dm \right) + \Delta P^T \left(\int \bar{P}\, dm \right) + \int \Delta P^T \Delta P\, dm$$

The first integral gives the total mass:

$$\int dm = m$$

The second and third integrals reduce to zero because they are integrating coordinates that are measured from the CG (as in Equation 5.73]:

$$\int \Delta P\, dm = \int \begin{pmatrix} 0 & -\Delta z & \Delta y \\ \Delta z & 0 & -\Delta x \\ -\Delta y & \Delta x & 0 \end{pmatrix} dm = \begin{pmatrix} 0 & 0 & 0 \\ 0 & 0 & 0 \\ 0 & 0 & 0 \end{pmatrix}$$

The final integral gives the total inertia matrix evaluated at the CG (which has the same form as Equation 5.79):

$$\int \Delta P^T \Delta P\, dm = \begin{pmatrix} \int (\Delta y^2 + \Delta z^2)\, dm & -\int \Delta z \Delta x\, dm & -\int \Delta x \Delta y\, dm \\ -\int \Delta z \Delta x\, dm & \int (\Delta z^2 + \Delta x^2)\, dm & -\int \Delta y \Delta z\, dm \\ -\int \Delta x \Delta y\, dm & -\int \Delta y \Delta z\, dm & \int (\Delta x^2 + \Delta y^2)\, dm \end{pmatrix}$$

So, for any object, the total inertia matrix J evaluated about any datum point is equal to the total inertia matrix J_G evaluated about the CG plus the inertia matrix $\bar{J}$ for the total mass (as an equivalent point mass at the CG) evaluated at the datum point. This is expressed as:

$$J = J_G + \bar{J} \tag{6.64}$$

where $\bar{J} = m\bar{P}^T\bar{P}$. This is known as the parallel axis theorem.

6.3.2 Transforming Mass Properties

Consider an object with mass properties as defined in Section 5.3.1. Recalling the discussion in Section 2.1, define a reference frame (or axis system) $\mathcal{F}^1$ for the object within which position vectors are specified for points inside the object boundary. The CG is defined by a position vector $\bar{P}^1$. The total kinetic energy is:

$$E = \frac{1}{2}\omega^{1T}J_G^1\omega^1 \tag{6.65}$$

where the inertia matrix J_G^1 is determined at the object CG using coordinates specified in $\mathcal{F}^1$.

If a reference frame $\mathcal{F}^0$ were introduced, then the angular velocity ω^1 would be re-expressed as:

$$\omega^1 = R^{10}\omega^0 \tag{6.66}$$

where R^{10} provides the coordinate transformation from $\mathcal{F}^0$ to $\mathcal{F}^1$. The kinetic energy would become:

$$E = \frac{1}{2}\omega^{0T}J_G^0\omega^0 \tag{6.67}$$

where

$$J_G^0 = R^{01}J_G^1R^{10} \tag{6.68}$$

Now, if the inertia matrix is determined at a point other than the object CG, then the parallel axis theorem (Equation 5.83) is applied in both reference frames:

$$J^0 = J_G^0 + \bar{J}^0 \qquad J^1 = J_G^1 + \bar{J}^1 \tag{6.69}$$

where

$$\bar{J}^0 = m\bar{P}^{0T}\bar{P}^0 \qquad \bar{J}^1 = m\bar{P}^{1T}\bar{P}^1$$

Combining Equations 6.68 and 6.69:

$$J^0 = R^{01}J_G^1R^{10} + \bar{J}^0 = R^{01}\left(J^1 - \bar{J}^1\right)R^{10} + \bar{J}^0$$

$$J^0 = R^{01}J^1R^{10} + \left(\bar{J}^0 - R^{01}\bar{J}^1R^{10}\right) \tag{6.70}$$

Also, if the origin of $\mathcal{F}^1$ is defined by a position vector r^{01} with respect to $\mathcal{F}^0$, then the CG position is transformed, as follows:

$$\bar{p}^0 = r^{01} + R^{01}\bar{p}^1 \tag{6.71}$$

This needs to be incorporated in the definition of the cross-product matrix $\bar{P}^0$.

6.3.3 Combining Mass Properties

Consider two objects with mass properties as defined previously. Let the first object have mass m_1, CG position $\bar{p}_1^1$, and inertia matrix J_1^1. Let the second object have mass m_2, CG position $\bar{p}_2^2$, and inertia matrix J_2^2. The subscripts give the index number for each object and the superscripts give the reference frame within the CG and the inertia are measured. To be consistent with previous sections, the reference frame for object #1 is $\mathcal{F}^1$ and the reference frame for object #2 is $\mathcal{F}^2$.

The combined mass m and CG position $\bar{p}^1$ are obtained, as follows:

$$m = m_1 + m_2 \tag{6.72}$$
$$m\bar{p}^1 = m_1\bar{p}_1^1 + m_2\bar{p}_2^1 \tag{6.73}$$

where

$$\bar{p}^1 = r^{12} + R^{12}\bar{p}^2 \tag{6.74}$$

The origin of $\mathcal{F}^2$ is defined by a position vector r^{12} with respect to $\mathcal{F}^1$. Note that the vector $\bar{p}^1$ determines the numerical values contained in $\bar{P}^1$ and $\bar{J}^1$.

The inertia matrix of object #2 is transformed from $\mathcal{F}^2$ to $\mathcal{F}^1$ according to Equation 5.89:

$$J_2^1 = R^{12}J_2^2R^{21} + \left(\bar{J}_2^1 - R^{12}\bar{J}_2^2R^{21}\right) \tag{6.75}$$

Thus, the combined inertia matrix J^1 is given by:

$$J^1 = J_1^1 + J_2^1 \tag{6.76}$$

6.3.4 Fuel Mass Distribution

Typically, fuel is distributed in designated wet zones within an aircraft structure, certainly in the wings and perhaps in the wing box and elsewhere in the fuselage (depending on the aircraft). The mass properties from all fuel cells are aggregated into mass properties for fuel tanks and thence the whole aircraft.

A simple oblong tank is shown in Figure 6.7. The enclosed volume is depicted as a point cloud fixed to a fixed-spaced rectangular grid. One point in this example corresponds with a 5 cm cube, such that eight points represent 1 l of fuel. A notional wing tank is shown in Figure 6.8, with a swept-back front spar and a slightly less swept-back rear spar. Setting the content to half-full and applying a 15° nose-up pitch angle, as in Figure 6.9, the point cloud distribution to the rear of the tank and has a fuel surface that is inclined with respect

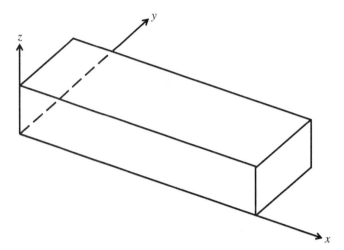

Figure 6.7 Point Cloud for an Oblong Tank.

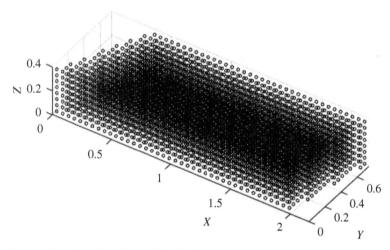

Figure 6.8 Point Cloud for a Wing Tank.

to the tank (albeit it appears like a staircase because the distribution is composed of discrete points).

Fuel height is measured perpendicular to the fuel surface, upwards from the lowest point in the tank (which depends on the orientation of the tank). Figure 6.10 gives an example of a height/volume graph, drawn from the results of numerical experiments ranging from 0 to 100% full and from −20° to +40° pitch. The top graph corresponds with −20° pitch, the bottom graph corresponds with +40° pitch and the straight-line graph corresponds with 0° pitch. The tank capacity here is 600 l but the actual volume occupied by a given mass

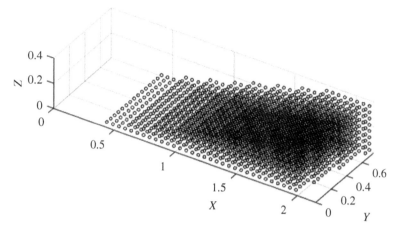

Figure 6.9 Point Cloud for a Wing Fuel Tank (50% Fill, 15-degree Pitch Angle).

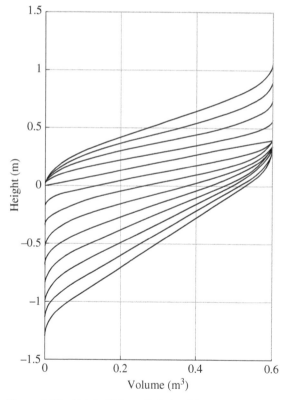

Figure 6.10 Height/Volume Relationship.

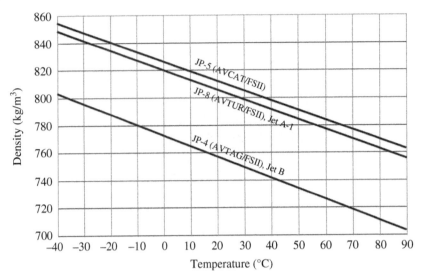

Figure 6.11 Fuel Density.

of fuel depends on fuel properties,[1] as shown in Figure 6.11. Note that re-fuelling at different airports will inevitably involve mixing fuel with different temperatures and different chemical compositions.

More elaborate models can be constructed for tank geometry, which give more accurate calculations of fuel contents. A tank can be specified as a set of bounding planes, each with a datum point and a normal vector pointing outwards (away from the interior of the tank). This information is processed in order to generate a faceted surface. Lines of intersection between planes are determined, which define the edges of the facets. Points of intersection between edges are determined, which define the vertices of the polyhedron.

This shape can then be subjected to volumetric analysis by introducing a cut-plane that represents the upper boundary of the fuel contained in the tank. This is a bounding plane that cuts the polyhedron into two segments, one wet and one dry. Fuel properties are calculated for the wet segment. Assuming that the tank is convex,[2] the wet segment is split into a set of space-filling tetrahedra for which mass properties are determined. These are then aggregated, as discussed earlier.

The simplest space-filling method is to choose any one of the vertices that define the space of the fuel 'solid'. From that position, the opposing facets are all visible, i.e. the interior of all facets that do not contain the chosen vertex. Each facet can be partitioned into $N - 2$ non-intersecting triangles (where N is the number of vertices on the boundary of that facet). New edges can be drawn from the chosen vertex to the corners of each triangle to produce a set of non-intersecting tetrahedra that span the entire volume.

Consider a solid tetrahedron with density ρ. If each object is defined by vertices (V_1, V_2, V_3, V_4), its centroid and its mass are defined by:

---

1 Source: Handbook of Aviation Fuel Properties, figure 3. [CRC Report 530, Coordinating Research Council Inc. 1983].
2 A convex shape has a bounding surface (or a set of bounding planes) that do not curve (or fold) inwards. Thus, any two points inside the shape can be joined by a straight line that is entirely contained within the shape.

$$\bar{P} = \frac{V_1 + V_2 + V_3 + V_4}{4} \qquad m = \frac{\rho}{6} |(V_1 - V_4) \cdot ((V_2 - V_4)) \times (V_3 - V_4)|$$

$$(6.77)$$

If the nth vertex V_n has coordinates (x_n, y_n, z_n) for $n = 1, ..., 4$, these calculations are performed as:

$$\bar{P} = \frac{1}{4} \begin{pmatrix} x_1 + x_2 + x_3 + x_4 \\ y_1 + y_2 + y_3 + y_4 \\ z_1 + z_2 + z_3 + z_4 \end{pmatrix} \qquad m = \frac{\rho}{6} \left| \det \begin{pmatrix} x_1 - x_4 & x_2 - x_4 & x_3 - x_4 \\ y_1 - y_4 & y_2 - y_4 & y_3 - y_4 \\ z_1 - z_4 & z_2 - z_4 & z_3 - z_4 \end{pmatrix} \right|$$

$$(6.78)$$

Following Tonon (2004), the inertia matrix is defined by:

$$J = \begin{pmatrix} J_x & -J_{xy} & -J_{zx} \\ -J_{xy} & J_y & -J_{yz} \\ -J_{zx} & -J_{yz} & J_z \end{pmatrix}$$

$$(6.79)$$

where

$$J_x = \frac{m}{10}(T_y + T_z) \qquad J_y = \frac{m}{10}(T_z + T_x) \qquad J_z = \frac{m}{10}(T_x + T_y)$$

and

$$J_{xy} = \frac{m}{20}(D_{xy} + S_{xy}) \qquad J_{yz} = \frac{m}{20}(D_{yz} + S_{yz}) \qquad J_{zx} = \frac{m}{20}(D_{zx} + S_{zx})$$

The components in these expressions can be generalised, as follows:

$$D_{uv} = \sum_{i=1}^{4} u_i v_i \qquad S_{uv} = \sum_{i=1}^{4}\sum_{j=1}^{4} u_i v_j \qquad T_u = \sum_{i=1}^{4}\sum_{j=i}^{4} u_i u_j$$

The generic coordinates u and v are replaced variously by x, y, and z, as appropriate.

This method allows a more accurate prediction of fuel tank contents and it can be applied for an entire fuel system, as illustrated in Figure 6.12. The wing tanks are typical of many

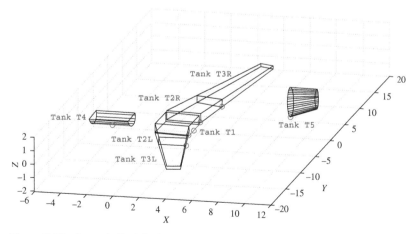

Figure 6.12 Example Fuel System.

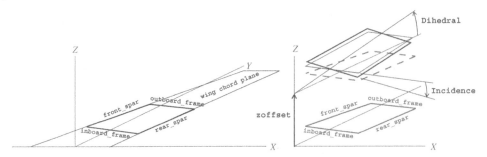

Figure 6.13 Schematic of Geometric Parameters.

large aircraft and a schematic is given in Figure 6.13, showing major geometric parameters for one wing tank (namely the planform boundary, vertical offset, wing dihedral, and wing incidence). Fuselage tanks are also included here as examples of what can be found in some configurations.

Figures 6.14 and 6.15 show the fuel distribution when the system is 50% full. This is shown at pitch angle of 0° and 20°, respectively. The shaded volume represents fuel and the small circles designated the fuel CG position in each tank. Note that the CG positions move rearwards when the aircraft pitches nose-up (exactly as expected).

The variation in CG position against fuel content is shown in Figure 6.16. This is called a fuel schedule in the United Kingdom and a burn curve in the United States. In this example, fuel is used in sequence from tanks T5 (Aft), T4 (Forward), T1 (Centre Wing), T2 (Inner Wing)), and finally T3 (Outer Wing). The total variation spans about 30cm. This can be reduced and re-shaped by changing the sequence. The variation due to pitch angle appears on the right-hand graph in Figure 6.16. In theory, it shows the extreme CG positions (fore

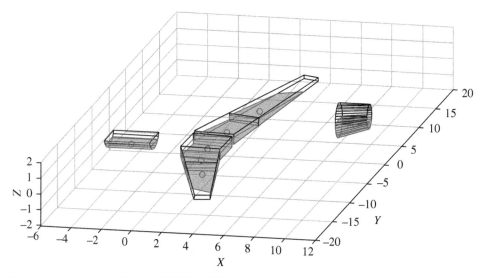

Figure 6.14 Fuel Distribution (50% Full, 0° Pitch).

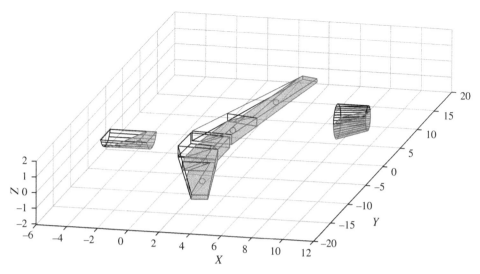

Figure 6.15 Fuel Distribution (50% Full, 20° Pitch).

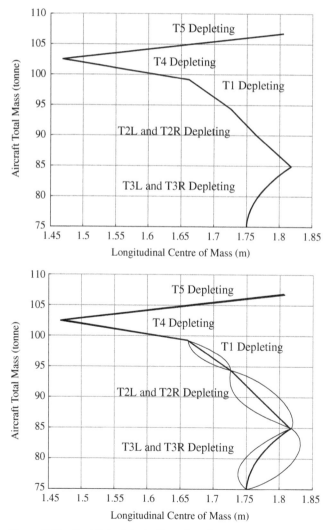

Figure 6.16 Fuel Schedule (or Burn Curve).

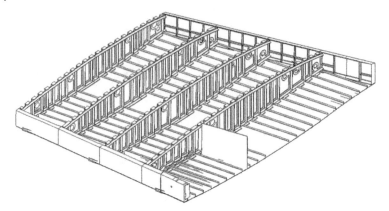

Figure 6.17   Typical Wing Structure.

and aft) for pitch angle in the range from −90° to +90°, as revealed by bulbous shapes super-imposed on the fuel schedule. Graphs of this type are called onion plots in the United Kingdom and potato plots in the United States.

In order to put this discussion into context, a typical wing structure is shown in Figure 6.17. This is what a real fuel tank will contain, namely the front spar and rear spar with structural frames running between them (although the inner-most frame is not shown). The tank floor is the lower wing surface, which is reinforced with stringers. There are five fuel cells (i.e. partitioned volumes), each of which needs to analysed separately. It will be noticed that the real geometry is not built from flat-sided boxes, as shown in the earlier (idealised) fuel system, but it can approximated by 'box' tanks for preliminary design calculations.

Bibliography

Anderson JD, 2011, Fundamentals of Aerodynamics, 5e, McGraw-Hill: New York.

Anderson RF, 1932, Charts for Determining the Pitching Moment of Tapered Wings with Sweepback and Twist, NACA Technical Report 483.

Anderson RF, 1936, Determination of the Characteristics of Tapered Wings, NACA Technical Report 572.

Babister AW, 1980, Aircraft Dynamic Stability and Response, Pergamon Press: Oxford.

Baldoino WM, Bodstein GCR, 2004, Comparative Analysis of the Extended Lifting-Line Theory to the Classical Lifting-Line Theory for Finite Wings, Proceedings ENCIT 2004, Rio de Janeiro, Brazil, 29 November–3 December 2004.

Bartel M, Young T, 2007, Simplified Thrust and SFC Calculations of Modern Two-Shaft Turbofan Engines for Preliminary Aircraft Design, 7th AIAA ATIO Conference, Belfast, Northern Ireland, September 2007. https://doi.org/10.2514/6.2007-7847

Bartel M, Young T, 2008, Simplified Thrust and Fuel Consumption Models for Modern Two-Shaft Turbofan Engines, J. Aircraft, 45(4). https://doi.org/10.2514/1.35589

Bennett BS, 1995, Simulation Fundamentals, Prentice Hall: London.

Bowring BR, 1976, Transformation from spatial to geodetic coordinates, Survey Review, 23(181): 323–327.

Breedveld PC, 1984a, A Bond Graph Algorithm to Determine the Equilibrium State of a System, J. Franklin Institute, 318(2), 71–75

Breedveld PC, 1984b, Physical Systems Theory in Terms of Bond Graphs, PhD Thesis, Universitiet Enschede, Netherlands.

van Broenink JF, 1997, Bond Graph Modeling in Modelica, European Simulation Symposium, Passau, Germany.

Brumbaugh RW, 1991, An Aircraft Model for the AIAA Controls Design Challenge, Paper AIAA-91-2631, Guidance and Control Conference, New Orleans, LA.

Cellier FE, 1991, Continuous System Modelling, Springer Verlag: Berlin.

Cohen H, Rogers GFC, Saravanamuttoo HIH, 1972, Gas Turbine Theory, 2e, Longman: London.

Dauphin-Tanguy G, Borne P, 1985, Order Reduction of Multi-time Scale Systems Using Bond Graphs, the Reciprocal System and the Singular Perturbation Method, J. Franklin Institute, 319(1/2), 157–171.

DeYoung J, Harper CW, 1955, Theoretical Symmetric Span Loading at Subsonic Speeds for Wings having Arbitrary Plan Form, NACA Report 921.

Diederich FW, 1952, A Simple Approximate Method for Calculating Spanwise Lift Distributions and Aerodynamic Influence Coefficients at Subsonic Speeds, NACA Technical Note 2751

van Dijk J, 1994, On the Role of Bond Graph Causality in Modelling Mechatronic Systems, PhD Thesis, Universitiet Twente, Netherlands.

Dommasch DO, Sherby SS, Connolly TF, 1956, Airplane Aerodynamics, Pitman: New York.

Eshelby M, 2000, Aircraft Performance: Theory and Practice, Butterworth-Heinemann: Oxford, UK.

Etkin B, 1972, Dynamics of Atmospheric Flight, John Wiley: New York.

Filippone A, 2012, Advanced Aircraft Flight Performance, Cambridge University Press: Cambridge, UK.

Flight Navigation Ltd, 1978, Three Dimensional BAC1-11/MONA RNAV Computer Model, Report FNL/7801/1.

Gantmacher FR, 1959, Theory of Matrices, Volume I, Chelsea: New York.

Gawthrop PJ, 1991, Bond Graphs: A Representation for Mechatronic Systems, Mechatronics, 1(2), 127–156.

Gawthrop PJ, 1998, Physical Interpretation of Inverse Dynamics using Bond Graphs, The Bond Graph Digest, 2(1). http://www.ece.arizone.edu/~cellier/bg_digest.html

Gawthrop PJ, Smith L, 1992, Causal Augmentation of Bond Graphs, J. Franklin Institute, 329(2), 291–303.

Gawthrop PJ, Smith L, 1996, Metamodelling: Bond Graphs and Dynamic Systems, Prentice Hall: London.

Göthert B, 1949, Airfoil Measurements in the DVL High-Speed Wing Tunnel (2.7 Metre Diameter), NACA Technical Memorandum 1240.

Harper CW, Maki RL, 1964, Review of the Stall Characteristics of Swept Wings, NASA Technical Note D-2373.

Harris CD, 1990, NASA Supercritical Airfoils, NASA Technical Paper 2969.

Heffley RK, Jewell WF, 1972, Aircraft Handling Qualities Data, NASA-CR-2144.

Hoak DE et al, 1960, USAF Stability and Control DATCOM, Air Force Wright Aeronautical Laboratories, TR-83-3048. [Revised 1978].

Hörner SF, 1951, Fluid Dynamic Drag, Otterbein Press: Ohio.

Howe D, 2000, Aircraft Conceptual Design Synthesis, Professional Engineering Publishing Limited: London, UK.

Ismail IH, 1991, Bhinder FS, Simulation of Aircraft Gas Turbines, J. Eng. Gas Turbines Power, 113(1), 95–99.

Jacobs EN, Ward KE, Pinkerton RM, 1932, The Characteristics of 78 Related Airfoil Sections from Tests in the Variable Density Wind Tunnel, NACA Report 460.

Karnopp DC, 1969, Power-conserving Transformations: Physical Interpretations and Applications using Bond Graphs, J. Franklin Institute, 288(3), 175–201.

Karnopp DC, 1978, Pseudo Bond Graphs for Thermal Energy Transport, ASME J. Dyn. Syst. Meas. Control, 100, 165–169.

Karnopp DC, 1979, State Variables and Pseudo Bond Graphs for Compressible Thermo-Fluid Systems, J. Dyn. Syst. Meas. Control 101(3): 201–204.

Karnopp DC, Margolis DL, Rosenberg RC, 1990, System Dynamics: A Unified Approach, 2e, John Wiley: New York.

Lange RH, 1945, A summary of Drag Results from Recent Langley Full Scale Tunnel Tests of Army and Navy Airplanes, NACA ARC L5A30 (declassified), Feb 1945.

Laitone EV, 1989, Lift-curve slope for finite-aspect-ratio wings, J. Aircraft, 26(8), 789–790.

Loftin LK, 1985, Quest for Performance: The Evolution of Modern Aircraft, NASA History Series Publication 468.

McCormick B, 1995, Aerodynamics, Aeronautics and Flight Mechanics, Wiley.

McRuer D, Ashkenas I, Graham D, 1973, Aircraft Dynamics and Automatic Control, Princeton University Press: Princeton, NJ.

MacFarlane AGJ, 1970, Dynamical System Models, GG Harrap: London.

Maccallum NRL, 1973, Effect of 'Bulk' Heat Transfer in Aircraft Gas Turbines on Compressor Surge Margins, IMechE Paper C36/73.

Maccallum NRL, 1976, Models for the Representation of Turbomachinery Blades During Temperature Transients, ASME Gas Turbine and Fluids Engineering Conference, New Orleans, LA.

Maccallum NRL, 1978, Thermal Influences in Gas Turbine Transients – Effects of Changes in Compressor Characteristics, ASME Gas Turbine Conference & Exhibition, San Diego, CA.

Maccallum NRL, Pilidis P, 1985, The Prediction of Surge Margins During Gas Turbine Transients, ASME Gas Turbine Conference & Exhibition, Houston, TX.

Mason WH, 1990, Analytic Models for Technology Integration in Aircraft Design, AIAA Paper 90-3262

Mattson SE, 1989, On Modelling of Differential/Algebraic Systems, Simulation, 24–32.

Nikravesh PE, 1988, Computer-Aided Analysis of Mechanical Systems, Prentice-Hall: Englewood Cliffs.

Niţă M, Scholz D, 2012, Estimating the Oswald Factor from Basic Aircraft Geometrical Parameters, Deutscher Luft- und Raumfahrtkongress, Estrel, Berlin, 10–12 September 2012.

Pamadi BN, 2004, Performance, Stability and Control of Airplanes, 2e, AIAA: Reston, VA.

Pamadi BN, 2015, Performance, Stability, Dynamics and Control of Airplanes, 3e, AIAA: Reston,VA, ISBN 978-1-62410-274-5.

Paynter KM, 1961, Analysis and Design of Engineering Systems, MIT Press: Boston, MA.

Rogers GFC, Mayhew YR, 1980, Engineering Thermodynamics and Heat Transfer, 3e, Longman: London.

Pelton JM, 1976, A Computer Model for the Calculation of Thermodynamic Properties of Working Fluids of a Gas Turbine Engine, Report AEDC-TR-76-15, Engine Test Facility, Arnold Engineering Development Center, Arnold Airforce Station.

Raymer DP, 1999, Aircraft Design: A Conceptual Approach, AIAA: Reston, VA.

Rolfe JM, Staples KJ (eds.), 1986, Flight Simulation, Cambridge University Press: Cambridge, UK.

Roskam J, 1989, Airplane Design, Volume 1: Preliminary Sizing of Airplanes, Design, Analysis and Research, DAR Corporation: Lawrence, KA.

Sforza PM, 2014, Commercial Airplane Design Principles, Elsevier Aerospace Engineering: Amsterdam, Netherlands.

Symon KR, 1971, Mechanics, Addison-Wesley: Reading, MA.

Takahashi T, German B, Shajanian A, Daskilewicz M, Donovan S, 2010, Zero Lift Drag and Drag Divergence Prediction for Finite Wings in Aircraft Conceptual Design. https://doi.org/10.2514/6.2010-846

Talay TA, 1975, Introduction to the Aerodynamics of Flight, NASA History Series Publication 367.

Teper GL, 1969, Aircraft Stability and Control Data, NASA-CR-96008.

Thoma J, 1975, Introduction to Bond Graphs and their Applications, Pergamon: Oxford.

Thoma J, 1990, Simulation by Bond Graphs, Springer-Verlag: Berlin.

Torenbeek E, 1982, Synthesis of Subsonic Airplane Design, Delft University Press: Delft, Netherlands.

Tonon F, 2004, Explicit Exact Formulas for the 3-D Tetrahedron Inertia Tensor in Terms of its Vertex Coordinates, J. Math. Stat., 1(1), 8–11

Wislicenus J, Daidzic NE, 2022, Estimation of Transport-category Jet Airplane Maximum Range and Cruising Airspeed in the Presence of Transonic Wave Drag, Aerospace J. 9(4), 1–28. https://doi.org/1s0.3390/aerospace9040192

Wellstead PE, 1979, Introduction to Physical System Modelling, Academic Press: London.

Zold T, 2012, Performance assessment of a hybrid electric-powered long-range commercial airliner. Technische Universitat Munchen

Index

Computational Modelling and Simulation of Aircraft and the Environment: Aircraft Dynamics,
First Edition, Volume II. Dominic J. Diston.
© 2024 John Wiley & Sons Ltd. Published 2024 by John Wiley & Sons Ltd.